1150
T3/20

Electron Waves and Resonances in Bounded Plasmas

Electron Waves and Resonances in Bounded Plasmas

PAUL E. VANDENPLAS
Professor,
Laboratoire de Physique des Plasmas,
Ecole Royale Militaire, Brussels, Belgium

1968

INTERSCIENCE PUBLISHERS
a division of John Wiley & Sons
London   New York   Sydney

Copyright © 1968 by John Wiley & Sons Ltd. All rights reserved. No part of this book may be reproduced by any means, nor transmitted, nor translated into a machine language without the written permission of the publisher.

Library of Congress catalog card No. 68–8597

SBN 470 90280 9

QC
718
.V29

Printed in Northern Ireland at The Universities Press

Preface

Some plasma physicists have expressed their concern about theoretical work which is not discussed in its relation to observational data. Most of the books on plasma physics published until now have not dealt extensively with the comparison between theory and experiments. This appears natural since many domains of plasma physics are in a state which renders such detailed comparison difficult, if not impossible.

The present monograph deals with one of the aspects of the fourth state of matter, namely electron waves in bounded plasmas. It is noteworthy that this area is characterized by the fact that detailed comparison between theory and observational data can indeed be carried out. As underlined in the introduction, a considerable (and sometimes very precise) body of information has been collected and this data on laboratory plasmas has been compared to theoretical work which takes boundary phenomena and boundary conditions into acount. A rewarding feature of this joint experimental and theoretical endeavour is that many phenomena are quantitatively explained. This general characteristic has been one of the principal motivations for undertaking the task of writing this book.

A remarkable fact is that the wave properties of bounded plasmas (and especially their resonances) differ considerably from those of infinite plasmas. Some of these properties are particularly striking when the characteristic dimension of the plasma is smaller than the vacuum wavelength. It is also becoming increasingly apparent that the behaviour exhibited at low power (linear phenomena) plays an important role in understanding the behaviour of the plasmas at high power. Such properties cannot therefore be neglected in the interpretation of plasma phenomena which occur in many laboratory devices. They also lead to interesting applications: new plasma diagnostic techniques and the possibility of enhanced radiation from a plasma coated antenna.

It is felt that physicists and engineers working in other aspects of plasma physics might be interested in a general exposition of results scattered in the literature and in an account which stresses the connexions between theory and experiments in one particular domain, as well as our understanding of some basic wave phenomena. Although the tract was written to be read in normal sequence, from beginning to end, care was taken to enable someone more interested in a particular chapter to understand it without being familiar with the book as a whole. To ensure this some important points have been repeated while a list of symbols and quite numerous cross-references are introduced.

It is always difficult to strike an exact balance between a general and clear exposition of the subject matter and the need for sufficient mathematical details. Since the book is also intended to serve as a convenient reference, sufficient steps in the mathematical derivation of important results have been included so as to render unnecessary, in many cases, recourse to the original literature. In a few instances, like chapter 4, the mathematics are rather heavy. This could not be avoided if the reader interested in the methods used, is to understand exactly the approximations performed and the deep reasons for the results obtained. On the other hand, the reader wanting to have a good picture without spending too much time will find, it is hoped, that the mathematical details can be cursorily read and that the text will draw his attention to the essential points. Some problems have been included in the text itself and quite a few more are nearly self-evident. In many cases these problems state a property, which the reader must not necessarily prove, and the physical implications of this property are then further discussed.

It is thought that the present book can serve to cover part of a general course on plasma physics, or on modern aspects of electromagnetic theory, or as a text for a course on specialized topics in plasma physics. The general level of the book is such that it can be used by advanced undergraduates or by graduate students in physics, engineering sciences and electrical engineering. In this connexion I would like to mention that preliminary notes of it have been used by Dr. J. Taillet, Head of the Saclay Plasma Laboratory and Visiting Professor at the Massachusetts Institute of Technology, in a seminar course given during the academic year 1965–1966 at the latter institution. He has been kind enough to tell me that it proved very suitable for an audience consisting of graduate students in physics and electrical engineering having only an elementary knowledge of plasma physics.

I had the opportunity of staying at Professor R. W. Gould's laboratory at the California Institute of Technology during part of the academic year 1959–1960 and then again in 1965–1966 at which time a substantial part of this book was written. It is a special pleasure to acknowledge my indebtedness to Roy Gould and to thank him for many enjoyable and stimulating discussions. I am grateful to the U.S. Office of Naval Research for providing partial support during my second visit to the California Institute of Technology. I would like to thank Drs. G. Barston, F. Blum, J. Taillet and Mr. J. J. Papier for comments on parts of the book and Mrs. R. Stratton for her very competent typing of the greater part of the manuscript. I also thank Miss J. Borremans for completing the typing of the manuscript.

Last, but not least, I extend my very sincere thanks to my collaborator, Dr. A. Messiaen, for having carefully read the entire manuscript and for help with the proofreading.

P. VANDENPLAS

June 1968

List of Notations and Symbols

b	inner radius of wall of plasma cylinder
B	perturbed magnetic induction
B$_0$	static magnetic induction
c	outer radius of wall of plasma cylinder
$-e$	electric charge of the electron
E	perturbed electric field strength
E$_0$, $\langle$**E**$\rangle$	static electric field strength
f	perturbed distribution function
F	distribution function
$\langle F \rangle$	equilibrium distribution function
$g(r)$	radial dependence of equilibrium density
I_{dis}	discharge current in plasma tube
J	current density
K_g	dielectric constant of glass
K_p	dielectric constant of cold plasma
m, m_e	mass of the electron
m_n	moment of order n
$N = \langle N \rangle + n$	particle electron density
$\langle N \rangle$	equilibrium electron density
n	perturbed electron density
$\langle N_0 \rangle$	equilibrium density on the axis of a non-uniform plasma column
$n = 1$	dipolar mode
$n = 2$	quadrupolar mode
$n = 3$	hexapolar mode
p, p_e	perturbed electron pressure
P, P_e	total electron pressure
Q	Q-factor
$r_D = \sqrt{(KT_e/m_e\omega_e^2)}$	Debye length
$r_\omega = \sqrt{(KT_e/m_e\omega^2)}$	
$1\,\text{T} = 10^4\,\text{G}$	1 Tesla $= 10^4$ Gauss
T, T_e	electron temperature
u	average perturbed electron velocity
v$_d$	drift velocity of the plasma
$\beta = \omega_c/\omega$	normalized static magnetic field
$\gamma = p/nKT$	

List of Symbols

ε_0	vacuum permittivity
$\varepsilon_p = \varepsilon_0(1 - \omega_e^2/\omega^2)$	scalar equivalent permittivity of electron plasma
$\boldsymbol{\epsilon}_p$	tensorial equivalent permittivity of electron plasma
μ_0	vacuum permeability
μ_e	electron mobility
ν, ν_e	electron collision frequency
σ	high frequency conductivity
ω	angular frequency of electromagnetic wave
$\omega_c = eB_0/m_e$	electron cyclotron frequency (angular)
$\omega_e = \left(\dfrac{e^2 \langle N \rangle}{\varepsilon_0 m_e}\right)^{\frac{1}{2}}$	electron plasma frequency (angular)
$(\omega_e^2)_{\text{Av}}$	radial average of ω_e^2 in a non-uniform plasma
$\omega_i = \left(\dfrac{e^2 \langle N \rangle}{\varepsilon_0 m_i}\right)^{\frac{1}{2}}$	ion plasma frequency (angular)
$\Omega^2 = \omega_e^2/\omega^2$	
$\Omega^2_{\text{Av}} = (\omega_e^2)_{\text{Av}}/\omega^2$	

Contents

Chapter 1 General Theoretical Methods and Experimental Techniques

1.1	Introduction	1
1.2	Equivalent permittivity of a plasma	3
1.3	Quasi-static approximation	8
1.4	Cold plasma cylinder in the quasi-static approximation	10
1.5	Oscillations and variational procedures in non-uniform cold plasmas	13
	a. Quasi-static oscillations in inhomogeneous cold plasmas	13
	b. Variational procedures	14
1.6	Moments method	17
	a. Dispersion relation	17
	b. Approximations made in the moments method	19
1.7	Boltzmann–Vlasov equation in cylindrical coordinates	23
1.8	Description of a warm non-uniform plasma with tensorial perturbed pressure	25
	a. Cylindrical case	26
	b. Tensorial generalization	28
1.9	Boundary conditions	28
	a. Cold plasmas	29
	b. Hot plasmas	29
1.10	Set-ups for studying wave-scattering by plasma columns	30
	a. Method of measurement of a reflexion factor in free space	31
	b. Detailed resonance spectrum exhibited by a plasma column	37
	c. Experiments in the waveguide	39
	d. Multipole electrode configurations	40
1.11	Measurement of electron plasma densities by cavity perturbation techniques	41

Chapter 2 **The Uniform Plasma Slab-Condenser System**

2.1 Theory 43
  a. Equations of the plasma slab-condenser system . 43
  b. Cold plasma ($T = 0$) 45
  c. Warm plasma ($T \neq 0$) 47
  d. Effect of collisions 53
  e. Summary 55
2.2 Experiments 57
2.3 High frequency plasmoids 60
2.4 Impedance of a plasma slab using Vlasov's equation 63
  a. Boundary conditions 64
  b. Steady state response to $E_0(t) = E_0 e^{i\omega t}$, $t > 0$. 66

Chapter 3 **The Hollow Cylindrical Plasma**

3.1 Theory 70
3.2 Comparison between experiments and theory . 73
3.3 Measurement of average plasma densities by the main resonances of cylindrical structures. . 79
3.4 Asymmetry effects in a hollow cylindrical plasma 80
  a. Theory 81
  b. Small eccentricity 85
  c. Comparison between theory and experiments . 86
  d. Conclusions. 89

Chapter 4 **Scattering of a Plane Electromagnetic Wave by a Plasma Column in Steady Magnetic Fields (Cold Plasma Approximation)**

4.1 Introduction 92
4.2 Anisotropy effects 95
  a. General formulation 95
  b. Theory and experiments with $\mathbf{B}_0$ parallel to the axis 99
  c. Theory with $\mathbf{B}_0$ perpendicular to the axis . . 104
4.3 Magnetically induced asymmetry effects . . 109
4.4 High-frequency effect due to axial velocity of the plasma column 116
  a. Theory 117
  b. Comparison with experimental data . . 122
  c. Final remarks 128

Chapter 5 Hot Non-uniform Plasma Column

5.1 Introduction 130
5.2 Perturbed scalar pressure approximation . . 132
   a. General formulation 132
   b. Parabolic density profile 134
   c. Tonks–Langmuir profile 142
   d. Detailed structure of the secondary spectrum . 147
   e. Damping effects 149
   f. Noise spectrum 152
   g. Asymmetry effects 154
   h. Axial propagation effects 158
   i. Conclusions. 160
5.3 Influence of steady axial magnetic fields . . 160
   a. Low magnetic fields ($\omega \gg \omega_c$). . . . 160
   b. High magnetic fields ($\omega = n\omega_c$; $n = 2, 3, 4, \ldots$) 161
   c. Continuity between low and high magnetic fields
      ($\omega_c \ll \omega$ to $\omega_c = \omega/2$) 165
5.4 Non-linear effects 166
   a. Non-linear temperature resonances . . . 166
   b. Non-linear resonances of a cold plasma . . 168
   c. Microwave scattering from density fluctuations
      resulting from temperature resonances . . 169
   d. Strong non-linear effects and resonantly sustained
      plasmas 171

Chapter 6 Metallic and Dielectric Resonance Probes. Plasma-dielectric Coated Antenna. General Considerations

6.1 Metallic resonance probe 173
6.2 High-frequency dielectric resonance probe . . 181
   a. Theory 182
   b. Experiments 187
6.3 Enhanced radiation from an antenna coated
   with dielectric and plasma 191
   a. Theory 192
   b. Experiments 196
   c. Non-linear behaviour 200
6.4 General considerations on resonances and anti-
   resonances of cold plasma systems . . . 201

Appendix 1 Volume Theorems Useful in Stating Mathematical Boundary Conditions

      1. Gradient theorem 206
      2. Divergence theorem 207
      3. Curl theorem 202
      4. Green's theorem 208

References 210

Index 217

CHAPTER 1

General Theoretical Methods and Experimental Techniques*

1.1 Introduction

Among the many fundamental problems of plasma physics, one that has received considerable attention is that of the interaction between plasmas and electromagnetic waves. Much work has been devoted to infinite and often homogeneous plasmas, since the theoretical treatment is then simpler. However, laboratory plasmas are finite and these have been studied actively during the last decade. The experimental techniques have been refined with the result that considerable and sometimes very precise observational data has been collected. The natural outcome is that quite a substantial amount of theoretical work on bounded plasmas has also been published. This work is frequently more complex and algebraically more tedious than that on infinite plasmas because boundary phenomena and boundary conditions between the plasma and the outside media have to be taken into account. A rewarding feature of this joint experimental and theoretical endeavour is that a certain number of phenomena have been quantitatively explained. This understanding is reflected by the development of new and rather precise measurement techniques: measurement of average plasma densities by means of the main resonance (or resonances) of plasma cylinders, the use of the metallic resonance probe and the dielectric two-wire resonance probe to determine electron plasma densities.

In this monograph we shall deal only with electron waves in bounded plasmas. We will further restrict the treatment to those cases for which the Poynting vector of the incoming electromagnetic field is perpendicular to the axis of the system (the characteristic dimension L of the plasma being smaller in most cases than the vacuum wavelength λ_0). The axis of the system is that direction in which the length of the plasma is not finite. In this domain of plasma physics there is generally quite good agreement between theory and experiments and accurate predictions can be made. However, it will be seen that in this area problems remain which are only partially understood.

The cases for which the Poynting vector is parallel to the axis, i.e. the axially propagating modes, have been extensively reviewed in the book by Allis, Buchsbaum and Bers[1]. We shall only indicate how the temperature

* Rationalized MKSA units are used throughout this book unless otherwise stated.

spectrum of chapter 5 can be connected to axially propagating modes by considering the resonance frequencies as the cut-off frequencies of certain propagating modes.

We now outline the contents of this tract:

(1) Chapter 1 is devoted to the derivation of some basic concepts, to a critical assessment of the precision of the moments method and to the exposition of some general methods and results.

(2) Chapter 2 deals, in the cold and hot plasma limits, with the model characterized by a collision-free plasma of uniform density and a variable region of vacuum between two condenser plates. The influence of collisions on the behaviour of the hot plasma is also investigated. Experimental results relative to the plasma slab-condenser system and to the high-frequency plasmoid slab are examined and compared to theoretical predictions. Finally, the integration of the Boltzmann–Vlasov equation is performed for the plasma slab and the results compared to those obtained by fluid theory.

(3) The cold hollow plasma cylinder is examined theoretically in chapter 3, together with the distinctive asymmetry effects caused by a lack of centring of the inner cylinder with respect to the outer one. Comparison with experimental data shows striking agreement. Examination of the results indicates that the resonant frequencies furnish quite a precise measurement of average plasma densities.

(4) The subject matter of chapter 4 is the magnetized cold plasma cylinder. We investigate the influence of an axial static magnetic field $\mathbf{B}_0$ on a cold homogeneous cylindrical plasma together with the anisotropy effects which appear in such a plasma when $\mathbf{B}_0$ lies in the cross-section of the cylinder. If there is an axial drift velocity of the electrons and if $\mathbf{B}_0$ lies in the cross-section of the cylinder, the resulting forces create a density inhomogeneity which is not radially symmetric. These asymmetric inhomogeneity effects are found to be much more important than the anisotropy effects and account for most of the observed scattering behaviour of the plasma. Finally, even in the absence of $\mathbf{B}_0$, the axial drift velocity of the plasma is shown to be a mechanism responsible for the coupling between the axial component E_z of the scattered field and the transverse incoming $\mathbf{E}$ field. This is a general effect due to the covariance of Maxwell's equations with respect to Lorentz transformations.

(5) The combined effects of temperature and non-uniformity on the resonant behaviour of a plasma column are investigated following the method outlined in chapter 1. The resonant frequencies of the temperature spectrum are found to be quite sensitive to the density

profile used and excellent agreement between theory and experiments is obtained by using the Tonks–Langmuir profile. The plot of the amplitude of the scattered field versus the equilibrium density is in striking agreement with observational data. The influence of steady magnetic fields on the temperature resonances is examined both experimentally and theoretically. Finally, interesting non-linear effects linked to the cold plasmas and to the temperature resonances are briefly mentioned.

(6) It is seen in chapter 6 that the model of chapter 2 furnishes a basis for a heuristic understanding of the metallic resonance probe. The resonance and the anti-resonance exhibited by this probe agree with a simple theory very well but the exact resonance value is linked to a rather precise knowledge of the sheath between the probe and the bulk of the plasma. The dielectric two-wire resonance probe is discussed both theoretically and experimentally. The resonance and anti-resonance displayed by this probe are quite insensitive to the sheath and the resonance value furnishes a precise value of electron plasma density. The effect of a steady magnetic field is also examined.

Several attempts have been made to predict the behaviour of an antenna coated with a plasma layer. These calculations omit, however, a salient feature of the experimental situation: namely the effect of the sheath between the plasma and the metallic antenna. It is shown that this sheath can play an essential role and can lead to a very strongly enhanced radiation by the antenna at certain plasma densities when the operating frequency is kept constant. This strongly enhanced radiation is fully confirmed by experiments.

Finally, general considerations are given which explain the characteristics of the resonant and anti-resonant behaviour of plasma resonators, be they plasma columns, the plasma slab-condenser system, the metallic or dielectric two-wire probes or the plasma-coated antenna.

1.2 Equivalent permittivity of a plasma

The interaction of an electromagnetic wave with a plasma disturbs the characteristic equilibrium quantities of the latter. We will show that it is possible to define an equivalent permittivity in the cold plasma approximation and that the plasma can then be considered as a frequency-dependent dielectric whose electromagnetic behaviour is described by Maxwell's source-free equations. We now consider certain macroscopic quantities of the plasma:

(i) particle electron density $\quad N_e = \langle N_e \rangle + n_e$
(ii) macroscopic electron velocity $\mathbf{U}_e = \langle \mathbf{U}_e \rangle + \mathbf{u}_e$
(iii) electron pressure $\quad P_e = \langle P_e \rangle + p_e$

where the time-independent or equilibrium quantities are indicated between brackets and the average perturbed quantities in small letters. Average here means, average over the different velocities of the particles considered and has nothing to do with a time average. Similar quantities are defined for the ions and are noted by a subscript i.

We make the following assumptions:

(1) The perturbations are small enough to neglect the product of perturbed quantities, i.e. we have a linearized theory.
(2) The perturbations have an $e^{-i\omega t}$ time-dependence where ω is the angular frequency of the electromagnetic field.
(3) Gravity forces are neglected.

The interaction between the plasma and the perturbing electromagnetic field is described macroscopically through Maxwell's equations, together with the hydrodynamic momentum equations for electrons and ions. The electric field strength in the plasma is denoted by $\mathbf{E}_0 + \mathbf{E}$ where $\mathbf{E}_0$ is the time-independent part and $\mathbf{E}$ is the $e^{-i\omega t}$ time-dependent part. Similarly, we denote the magnetic induction by $\mathbf{B}_0 + \mathbf{B}$ and the current density by $\mathbf{j}_0 + \mathbf{j}$. Maxwell's equations for the HF components are then

$$\text{curl } \mathbf{H} = \mathbf{j} - i\omega\epsilon_0 \mathbf{E} \tag{1-1}$$

$$\text{curl } \mathbf{E} = i\omega \mathbf{B} \tag{1-2}$$

$$\text{div } \mathbf{B} = 0 \tag{1-3}$$

$$\text{div } \mathbf{E} = \rho/\epsilon_0 \tag{1-4}$$

with $\mathbf{B} = \mu_0 \mathbf{H}$, ϵ_0 and μ_0 being the permittivity and permeability of vacuum respectively. The electric charges are considered as moving in vacuum; the charge density ρ and the current density $\mathbf{j}$ are defined through the plasma quantities as follows:

$$\rho = -en_e + Zen_i \tag{1-5}$$

$$\mathbf{j} = -e[\langle N_e \rangle \mathbf{u}_e + \langle \mathbf{U}_e \rangle n_e] + Ze[\langle N_i \rangle \mathbf{u}_i + \langle \mathbf{U}_i \rangle n_i] \tag{1-6}$$

where e is the absolute value of the charge of the electron and Ze is the charge of the ions. The identity div curl $\equiv 0$ applied to Maxwell's equations leads to the continuity equation

$$\nabla \mathbf{j} = i\omega e(Zn_i - n_e) \tag{1-7}$$

Equivalent Permittivity of a Plasma

The linearized hydrodynamic equation (i.e. where the product of perturbed quantities is neglected) for the transport of electron momentum is given, (see for example reference 2), by

$$-i\omega \mathbf{u}_e + [\langle \mathbf{U}_e \rangle . \nabla] \mathbf{u}_e = -\frac{e}{m_e}[\mathbf{E} + \langle \mathbf{U}_e \rangle \times \mathbf{B} + \mathbf{u}_e \times \mathbf{B}_0]$$
$$-\frac{\nabla p_e}{m_e \langle N_e \rangle} - \frac{e n_e \mathbf{E}_0}{m_e \langle N_e \rangle} + \frac{\mathbf{\Gamma}_e}{m_e \langle N_e \rangle} \quad (1\text{-}8)$$

with m_e being the mass of the electron and where $\mathbf{\Gamma}_e$ represents the change of momentum per unit time and per unit volume due to the collisions of electrons with ions and neutrals. $\mathbf{\Gamma}_e$ is thus a damping term of the hydrodynamic equation. A simplified expression for $\mathbf{\Gamma}_e$ is the following:

$$\mathbf{\Gamma}_e = -\langle N_e \rangle \nu_{ei} m_e (\mathbf{u}_e - \mathbf{u}_i) - \langle N_e \rangle \nu_{en} m_e (\mathbf{u}_e - \mathbf{u}_n) \quad (1\text{-}9)$$

where ν_{ei} and ν_{en} are the collision frequencies between electrons and ions and between electrons and neutral particles respectively, $\mathbf{u}_n$ being the average velocity of the neutrals which is equal to zero in many cases. Equation (1-9) must be considered as a simplified phenomenological definition of the collision frequencies. In all the cases that will be considered in this book $|\mathbf{u}_e| \gg |\mathbf{u}_i|$, then (1-9) becomes

$$\mathbf{\Gamma}_e = -\langle N_e \rangle m_e \nu \mathbf{u}_e \quad (1\text{-}10)$$

For the ions we have an equation exactly similar to (1-8), namely

$$-i\omega \mathbf{u}_i + [\langle \mathbf{U}_i \rangle . \nabla] \mathbf{u}_i = \frac{Ze}{m_i}[\mathbf{E} + \langle \mathbf{U}_i \rangle \times \mathbf{B} + \mathbf{u}_i \times \mathbf{B}_0]$$
$$-\frac{\nabla p_i}{m_i \langle N_i \rangle} + \frac{Z e n_i \mathbf{E}_0}{m_i \langle N_i \rangle} + \frac{\mathbf{\Gamma}_i}{m_i \langle N_i \rangle} \quad (1\text{-}11)$$

Plasmas are always characterized by $m_i \gg m_e$. For a hydrogen plasma $m_i/m_e \simeq 2 \times 10^3$ and for a mercury plasma $m_i/m_e \simeq 3.5 \times 10^5$. In the limit $m_i/m_e \to \infty$, $\mathbf{u}_i$ and n_i tend to zero, ($\mathbf{u}_i$ and n_i are linked by the ion continuity equation) and we have no ion motion. This is our fourth assumption:

(4) The motion of the ions is neglected.

In that limit the hydrodynamic behaviour of the plasma is described by equation (1-8) but a relation connecting p_e and n_e is necessary in order to make the term ∇p_e explicit. Assuming a scalar electron pressure, we have

$$p_e = \gamma n_e K T_e \quad (1\text{-}12)$$

where T_e is the electron thermodynamic temperature, K is Boltzmann's constant and γ is a coefficient depending on the perturbation law chosen.

The validity of such an assumption will be assessed later (see sections 1.6 and 1.8).

Using (1-10) and (1-12), equation (1-8) can, in the absence of an electrostatic field $\mathbf{E}_0$, be written as follows

$$-i\omega \mathbf{u}_e + [\langle \mathbf{U}_e \rangle \cdot \nabla] \mathbf{u}_e = -\frac{e}{m_e}[\mathbf{E} + \langle \mathbf{U}_e \rangle \times \mathbf{B} + \mathbf{u}_e \times \mathbf{B}_0]$$
$$-\frac{\gamma K T_e \nabla n_e}{m_e \langle N_e \rangle} - \nu \mathbf{u}_e \quad (1\text{-}13)$$

If we want to describe the plasma as a dielectric medium characterized by an equivalent permittivity, we have to establish a linear relation between $\mathbf{j}$ and $\mathbf{E}$ of the form

$$\mathbf{j} = \boldsymbol{\sigma} \cdot \mathbf{E} \quad (1\text{-}14)$$

where $\boldsymbol{\sigma}$ is the equivalent conductivity tensor of the plasma. Note that this high frequency conductivity is defined in a purely formal way; indeed, a finite conductivity is obtained even when there are no collisions ($\nu = 0$). It should also be noted that this is not the only conductivity that is defined in a plasma. To obtain equation (1-14) we make two further assumptions:

(5) There is no electron drift velocity, $\langle \mathbf{U}_e \rangle = 0$
(6) $T_e = 0$; this is the cold plasma approximation.

This last assumption is not absolutely necessary to define an equivalent conductivity tensor. Indeed, the continuity equation enables one to express ∇n_e as a function of $\mathbf{j}$ through a formula in which the components of the wave vector $\mathbf{k}$ in the plasma intervene. It is difficult, however, to express these components as a function of ω and the $\boldsymbol{\sigma}$ obtained is very cumbersome to use[20,3].

(a) $\mathbf{B}_0 = 0$. With no applied steady magnetic field, equations (1-6) and (1-13) lead to

$$\mathbf{u}_e = \frac{-\mathbf{j}}{e \langle N_e \rangle} = -\frac{e\mathbf{E}}{m_e(\nu - i\omega)} \quad (1\text{-}15)$$

The high frequency conductivity of the plasma—also known as the Lorentz conductivity—is then scalar and given by

$$\sigma = \frac{e^2 \langle N_e \rangle}{m_e(\nu - i\omega)} = \frac{\epsilon_0 \omega_e^2}{(\nu - i\omega)} \quad (1\text{-}16)$$

where

$$\omega_e^2 = \frac{e^2 \langle N_e \rangle}{\epsilon_0 m_e} \quad (1\text{-}17)$$

ω_e is called the electron plasma angular frequency and is the natural frequency of oscillation of an infinite plasma in the approximation made. As

Equivalent Permittivity of a Plasma

will be seen later, it is also the cut-off frequency for the propagation of electromagnetic waves in a plasma.

Introducing $\mathbf{j} = \boldsymbol{\sigma} \cdot \mathbf{E}$ in equation (1-1), we obtain the equivalent permittivity ϵ_p of the plasma defined through curl $\mathbf{H} = -i\omega\epsilon_p\mathbf{E}$:

$$\epsilon_p = \epsilon_0\left(1 - \frac{\omega_e^2}{\omega^2 + \nu^2} + \frac{i\nu}{\omega}\frac{\omega_e^2}{\omega^2 + \nu^2}\right) \tag{1-18}$$

If $\nu/\omega \ll 1$, i.e. if the frequency of the wave is much greater than the collision frequency (this will be the case in many of the situations investigated in this book), we obtain

$$\epsilon_p = \epsilon_0(1 - \Omega^2) \tag{1-19}$$

after introducing the reduced variable $\Omega = \omega_e/\omega$.

(b) $\mathbf{B}_0 \neq 0$. We first express this permittivity in any orthogonal curvilinear coordinate system characterized by the unit vectors $\mathbf{l}_1$, $\mathbf{l}_2$ and $\mathbf{l}_3$. The static magnetic induction $\mathbf{B}_0$ is characterized by the components B_{01}, B_{02} and B_{03}. Equation (1-13) gives in the absence of collisions:

$$\begin{bmatrix} E_1 \\ E_2 \\ E_3 \end{bmatrix} = \frac{1}{\epsilon_0\omega_e^2}\begin{bmatrix} -i\omega & \frac{e}{m_e}B_{03} & -\frac{e}{m_e}B_{02} \\ -\frac{e}{m_e}B_{03} & -i\omega & \frac{e}{m_e}B_{01} \\ \frac{e}{m_e}B_{02} & -\frac{e}{m_e}B_{01} & -i\omega \end{bmatrix}\begin{bmatrix} j_1 \\ j_2 \\ j_3 \end{bmatrix} \tag{1-20}$$

or $\mathbf{E} = \boldsymbol{\sigma}^{-1} \cdot \mathbf{j}$.

Introducing $\mathbf{j} = \boldsymbol{\sigma} \cdot \mathbf{E}$ in equation (1-1), we obtain the tensorial permittivity $\boldsymbol{\epsilon}_p$:

$$\boldsymbol{\epsilon}_p = \epsilon_0\begin{bmatrix} 1 - \Omega^2\left(\dfrac{1-\beta_1^2}{1-\beta^2}\right) & \Omega^2\left(\dfrac{\beta_1\beta_2 + i\beta_3}{1-\beta^2}\right) & \Omega^2\left(\dfrac{\beta_1\beta_3 - i\beta_2}{1-\beta^2}\right) \\ \Omega^2\left(\dfrac{\beta_1\beta_2 - i\beta_3}{1-\beta^2}\right) & 1 - \Omega^2\left(\dfrac{1-\beta_2^2}{1-\beta^2}\right) & \Omega^2\left(\dfrac{\beta_2\beta_3 + i\beta_1}{1-\beta^2}\right) \\ \Omega^2\left(\dfrac{\beta_1\beta_3 + i\beta_2}{1-\beta^2}\right) & \Omega^2\left(\dfrac{\beta_2\beta_3 - i\beta_1}{1-\beta^2}\right) & 1 - \Omega^2\left(\dfrac{1-\beta_3^2}{1-\beta^2}\right) \end{bmatrix} \tag{1-21}$$

with $B_g = \omega m_e/e$, $\beta_1 = B_{01}/B_g$, $\beta_2 = B_{02}/B_g$, $\beta_3 = B_{03}/B_g$, $\beta = B_0/B_g = (\beta_1^2 + \beta_2^2 + \beta_3^2)^{\frac{1}{2}} = \omega_c/\omega$, and ω_c is the cyclotron frequency corresponding to B_0. In a Cartesian coordinate system $\boldsymbol{\epsilon}_p$ is also given by (1-21). If $\mathbf{B}_0$ is parallel to an axis, say the z-axis, the Hermitian tensor $\boldsymbol{\epsilon}_p$

becomes

$$\boldsymbol{\epsilon}_p = \epsilon_0 \begin{bmatrix} 1 - \dfrac{\Omega^2}{1-\beta^2} & \dfrac{i\beta\Omega^2}{1-\beta^2} & 0 \\ -\dfrac{i\beta\Omega^2}{1-\beta^2} & 1 - \dfrac{\Omega^2}{1-\beta^2} & 0 \\ 0 & 0 & 1-\Omega^2 \end{bmatrix} \quad (1\text{-}22)$$

If $\beta = 0$ formula (1-22) reduces of course to the scalar form (1-19).

Whereas the propagation of an electromagnetic wave in a plasma is a self consistent field problem generally described by equations (1-1) to (1-6) together with (1-8) and (1-9), the introduction of an equivalent permittivity made possible by the supplementary assumptions (4), (5) and (6), reduces the problem to the solution of Maxwell's equations in a dielectric medium characterized by $\boldsymbol{\epsilon}_p$ and μ_0:

$$\operatorname{curl} \frac{\mathbf{B}}{\mu_0} = -i\omega \boldsymbol{\epsilon}_p \cdot \mathbf{E} \quad (1\text{-}23)$$

$$\operatorname{curl} \mathbf{E} = i\omega \mathbf{B} \quad (1\text{-}24)$$

$$\operatorname{div} \mathbf{B} = 0 \quad (1\text{-}25)$$

$$\operatorname{div} \boldsymbol{\epsilon}_p \cdot \mathbf{E} = 0 \quad (1\text{-}26)$$

Note that (1-26) results from (1-23) through the identity div curl $\equiv 0$.

When a plasma is described in this approximation, the difficulty lies in the fact that the equivalent dielectric is generally inhomogeneous and anisotropic. Remember, however, that $\mathbf{E}_0$ and/or $\langle \mathbf{U}_e \rangle$ are present when the plasma is inhomogeneous and that they have not been taken into account in the derivation of $\boldsymbol{\epsilon}_p$. This problem of inhomogeneous anisotropic plasmas is treated for plane waves and indefinite media in a very thorough way by Ginzburg[5]. The anisotropy and asymmetrical inhomogeneity effects in a cylinder, i.e. in a bounded plasma, will be examined through the equivalent permittivity approach in chapter 4.

1.3 Quasi-static approximation

Many authors make the quasi-static approximation which corresponds to setting $-\partial \mathbf{B}/\partial t = i\omega \mathbf{B}$ equal to zero in equation (1-24). The identity curl grad $\equiv 0$ leads then to $\mathbf{E} = -\nabla \phi$ and the electric field derives from a potential function ϕ. This approximation is often very useful and leads to precise results when it is warranted. However, it is not always completely clear when this approximation can be made. A simple way of assessing it is the following[21].

Quasi-Static Approximation

Consider a cold uniform plasma with $B_0 = 0$. Equation (1-26) gives, in the quasi-static approximation

$$\epsilon_p \nabla^2 \phi = 0 \qquad (1\text{-}27)$$

and one must solve the Laplace equation for the electric potential when $\epsilon_p \neq 0$. Equations (1-23) and (1-24) lead, in Cartesian coordinates, to the Helmholtz equation

$$(\nabla^2 + k_p^2)\mathbf{E} = 0 \quad \text{with} \quad k_p^2 = \epsilon_p \mu_0 \omega^2 \qquad (1\text{-}28)$$

But if $\epsilon_p \neq 0$, the gradient of equation (1-27) gives

$$\nabla^2 \mathbf{E} = 0 \qquad (1\text{-}29)$$

The quasi-static approximation will thus be warranted when the solutions of equations (1-28) and (1-29) are asymptotically identical in the region of interest. Consider, e.g. a case when $\mathbf{E}$ depends on only one coordinate, say x, and examine a given component, say E. Choosing the origin of the x-axis at the centre of our system, the exact solution of equation (1-28) is

$$E_H = E_0 \cos k_p x + \left(\frac{\partial E}{\partial x}\right)_0 \frac{\sin k_p x}{k_p} \qquad (1\text{-}30)$$

where the subscript zero indicates that the quantities are evaluated at $x = 0$. The solution E_L of equation (1-29) is

$$E_L = E_0 + \left(\frac{\partial E}{\partial x}\right)_0 x \qquad (1\text{-}31)$$

The series expansion of (1-30) shows that $E_L \simeq E_H$ when $|k_p x|^2 \ll 1$. The criterion for using the quasi-static approximation in such a plasma is thus

$$|\epsilon_p \mu_0 \omega^2 L^2| \ll 1 \qquad (1\text{-}32)$$

where L is a characteristic dimension of the system. We must immediately indicate that the quasi-static approximation does not coincide with the assumption of purely longitudinal waves; this point is discussed at some length in section 2.1a.

One must, however, be careful in using the quasi-static approach. It can be shown that in cylindrical coordinates r, θ, z the axially symmetric solution (no θ-angular dependence) cannot be obtained from ϕ[36]. This is due to the fact that the axially symmetric potential obtained by solving the Laplace equation is $\phi = A + B \ln r$ and that this ϕ is singular at $r = 0$ and $r = \infty$: such a solution does not, evidently, have a proper behaviour on the axis and at infinity.

In the case of a hot plasma, the approximation made when using the quasi-static approximation can be estimated in the following way[62].

Consider a hot uniform plasma which is described by Maxwell's equations and by electric charges moving in vacuum. One must solve

$$(\nabla^2 + k_0^2)\mathbf{H} = -\text{curl } \mathbf{j} \tag{1-33}$$

where k_0 is the magnitude of the wave vector in vacuum and is given by $k_0^2 = \epsilon_0\mu_0\omega^2$. The current $\mathbf{j} = -e\langle N_e\rangle\mathbf{u}_e$ is given by the corresponding hydrodynamic equation and we have a self-consistent problem. When the quasi-approximation $\mathbf{E} = -\nabla\phi$ is made, one solves in fact $\nabla^2\mathbf{H} = -\text{curl } \mathbf{j}$ instead of the correct equation (1-33). This is warranted if $k_0^2 L^2 \ll 1$ where L is again a characteristic dimension of the plasma.

1.4 Cold plasma cylinder in the quasi-static approximation

We briefly investigate the behaviour of a uniform cold plasma column of radius b placed in vacuum when it is subjected to an incoming plane wave of unit amplitude. The electric field strength **E** and Poynting vector **S** of the incoming wave are perpendicular to the axis of the column. This whole problem is treated in a general fashion in chapter 3 for the case of a hollow cylinder (see figure 3.1) and it will be seen there, that the case of the plasma column is re-obtained by letting the innermost radius a of the hollow cylinder tend to zero.

The uniform cold plasma column is simple to handle and is of historical significance, since Tonks[63] showed as early as 1931 that is exhibits resonance oscillations at an angular frequency ω equal to $\omega_e/\sqrt{2}$. This result was rederived by Herlofson[64] in his detailed study of a plasma column in connexion with resonant scattering from ionized meteor trails.

We treat this problem in the quasi-static approximation. The solutions of Laplace's equation in cylindrical coordinates having the correct behaviour at $r = 0$ and $r = \infty$ are

$$\phi_1 = B_n r^n \cos n\theta; \quad r < b \tag{1-34}$$

$$\phi_2 = G_n r^{-n} \cos n\theta + r^n \cos n\theta; \quad r > b \tag{1-35}$$

where the last term in equation (1-35) represents the exciting or incoming field. In fact, as the quasi-static approximation implies $b \ll \lambda_0$, this means that the field seen by the column is practically uniform, i.e. the excitation is preponderantly dipolar ($n = 1$). For more details on this, see section 3.1.

The constants B_n and G_n are determined by the boundary conditions. As the problem at hand is formally identical to the electromagnetic dielectric problem (section 1.2), the boundary conditions at the surface of discontinuity between plasma and vacuum are clearly the continuity of the

Cold Plasma Cylinder in the Quasi-Static Approximation

tangential components of electric field strength and the continuity of the normal components of the electric displacement. This yields

$$G_n = b^{2n} \frac{\epsilon_0 - \epsilon_p}{\epsilon_0 + \epsilon_p} \tag{1-36}$$

When $\epsilon_p = -\epsilon_0$, i.e. when $\omega = \omega_e/\sqrt{2}$, G_n is infinite and resonance scattering is observed. That the maximum of G_n corresponds to infinity is due to the fact that the quasi-static approximation neglects radiation damping (see problem following equation (3-16) in section 3.1) and that the plasma was assumed collision free. When $\epsilon_p = \epsilon_0$ there is no plasma and $G_n = 0$ naturally reflects the fact that then there is no scattering.

If no exciting field is present, $\omega = \omega_e/\sqrt{2}$ is clearly the compatibility condition, that is, the condition to be fulfilled in order that a solution exists with B_n and G_n not trivially zero. The frequency $\omega_e/\sqrt{2}$ is the natural oscillation frequency of a plasma column described in this approximation.

Figures (3.4a) and (3.4b) of chapter 3 show the comparison between the theoretical curve of $|G_1|$ versus ω_e^2 and the corresponding experimental curve. (Note that the behaviour is slightly altered by the presence of a glass wall, i.e. $|G_1|$ does not tend to zero when $\omega_e^2 \to 0$.) This comparison clearly indicates that the theoretical description of a real plasma (which is non-uniform and hot) in the uniform cold plasma limit conveys valuable (and, maybe rather surprisingly, fairly accurate) information concerning the overall high-frequency response of such a plasma.

One is interested in knowing how the non-uniformity of the plasma will affect the previous results in the cold plasma limit. This can be done by using an inhomogeneous ϵ_p. It must be remembered, as remarked above, that in addition to the six assumptions underlined, this involves also the neglect of the term containing the static field $\mathbf{E}_0$; it is this field which balances the non-uniformity. As an example, we consider the non-uniform plasma column in vacuum. With $\phi = R(r) \cos n\theta$, equation (1-26) leads to

$$\frac{d^2 R}{dr^2} + \left(\frac{1}{r} + \frac{1}{\epsilon_p} \frac{d\epsilon_p}{dr} \right) \frac{dR}{dr} - \frac{n^2}{r^2} R = 0 \tag{1-37}$$

Assuming a parabolic electron density profile given by

$$\langle N \rangle = \langle N_0 \rangle \left[1 - \alpha \left(\frac{r}{b} \right)^2 \right]; \quad r < b \tag{1-38}$$

the resonance frequency of the column is sought following the procedure described above. The square of the *average* plasma frequency is defined by

$$(\omega_e^2)_{\text{Av}} = \omega_{e0}^2 \int_0^b \left[1 - \alpha \left(\frac{r}{b} \right)^2 \right] \frac{2r \, dr}{b^2} = \omega_{e0}^2 \left(1 - \frac{\alpha}{2} \right) \tag{1-39}$$

where ω_{eo}^2 corresponds to $\langle N_0 \rangle$. The results found[67] are summarized in figure 1.1. The method of solution is not valid in the shaded area, since a solution in this region would have the resonant frequency equal to the local plasma frequency at some radius in the column. For this radius, $\epsilon_p = 0$, and we have a singular point in equation (1-37); the solution is only valid for $\epsilon_p < 0$.

When $\alpha < 0\cdot 68$, the resonant frequency ω_R is approximately given by $\sqrt{((\omega_e^2)_{\mathrm{Av}}/2)}$, i.e. we have the same result as for a uniform plasma with an

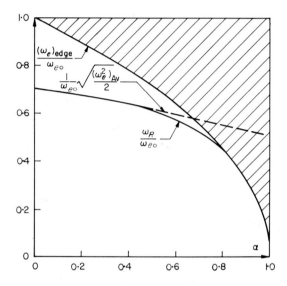

Figure 1.1 Resonant frequency ω_R of a non-uniform plasma column. The method of solution is not valid in the shaded area.

ω_e equal to $(\omega_e)_{\mathrm{Av}}$. When $\alpha > 0\cdot 68$, the resonant frequency is given approximately by the plasma frequency at the edge of the column; the fundamental mathematical reason for this will appear in the following section and it should here be emphasized that the use of an inhomogeneous ϵ_p has thus rather severe restrictions in a bounded plasma. It will, however, be seen in chapter 5 that when a hot non-uniform plasma column is studied, the main resonance (i.e. the resonance obtained in the cold plasma limit) is very adequately described by using a cold uniform plasma with a density equal to the average density in the actual non-uniform column if the Debye radius is small with respect to the characteristic dimension b of the system.

1.5 Oscillations and variational procedures in non-uniform cold plasmas

a. Quasi-static oscillations in inhomogeneous cold plasmas

The oscillations of cold inhomogeneous plasmas have been investigated in connexion with the problem of the eigenfrequencies of such a plasma. Several authors[64,25] have been concerned with the singular nature of the points where $\epsilon_p = 0$, i.e. where $\omega_e = \omega$, and with the fact that the plasma resonates there with an infinite value of the electric field. The defects of the model in this region have been pointed out; it was remarked[21] that a finite cold plasma would have as many resonance frequencies as there are jump discontinuities in a step-like approximation of a continuous density distribution and that it was therefore expected that such a plasma would resonate over the continuous range of ω for which ω is equal to ω_e somewhere in the plasma. A systematic theoretical study of the small-amplitude electrostatic oscillations in cold inhomogeneous plasmas has been carried out by Barston in 1963[68,69,70] and we will give the summary of his findings.

Consider a cold non-uniform plasma *at rest**. As in the description by a non-uniform ϵ_p, the perturbed term proportional to the unperturbed electric field $\mathbf{E}_0$ is neglected. Barston's analysis leads to the following results.

Well-behaved solutions to the linearized partial differential equations describing the system are obtained by taking a Fourier integral superposition of the singular modes ω over their continuous spectrum. Characteristically, one finds that:

1. A finite plasma with an everywhere continuous and nowhere constant plasma frequency $\omega_e(\mathbf{r})$ (where $\mathbf{r}$ is the position vector) allows only singular modes and an entirely continuous spectrum. No dispersion relation exists.
2. If, for a collision-free plasma, $\omega_e^2(\mathbf{r})$ is continuous in a certain region D, that region contributes to the spectrum precisely those values of ω such that $\omega^2 = \omega_e^2(\mathbf{r})$ for some $\mathbf{r}$ in D.
3. A finite plasma with an everywhere continuous $\omega_e^2(\mathbf{r})$ will resonate in the presence of an applied field, oscillating with frequency ω precisely at those points $\mathbf{r}$ of the plasma where $\omega_e^2(\mathbf{r}) = \omega^2$.
4. The introduction of jump discontinuities into $\omega_e^2(\mathbf{r})$ is necessary for the existence of well-behaved modes and for the existence of a

---

* For cold non-uniform streaming plasmas, these results do not apply and one might perhaps expect less drastic restrictions. See, for example Problem 2 of reference 69.

discrete spectrum if ω_e^2 is nowhere constant. If $\omega_e^2(\mathbf{r})$ jumps discontinuously from a value ω_{e1}^2 to a value ω_{e2}^2 greater than ω_{e1}^2 upon crossing a surface S, and if $(\omega_{e1}^2, \omega_{e2}^2)$ and the range of $\omega_e^2(\mathbf{r})$ have no points in common, then a well-behaved mode will exist with a resonance frequency ω_R such that $\omega_{e1}^2 < \omega_R^2 < \omega_{e2}^2$.

This last point is well illustrated by the example treated at the end of the preceding section. The resonance frequency ω_R was always such that $0 < \omega_R^2 < (\omega_e^2)_{\text{edge}}$. Furthermore, it will be seen in chapter 5 that when the Debye radius is sufficiently small with respect to the dimensions of the plasma, the frequency of the main resonance of a hot non-uniform plasma cylinder is given accurately by the cold plasma calculation using a uniform density equal to the average one.

It is therefore clear that, in most instances, there is little practical value in considering an inhomogeneous cold plasma: either the correction is small (see the example of section 1.4 when $\alpha < 0.68$) and one does not know whether it is relevant or, if the density at the edge of the plasma becomes too small, one gets an incorrect answer because of the mathematical condition $\omega_R < (\omega_e)_{\text{edge}}$ imposed by the very model used. On the other hand, the effect of a rather small asymmetrical non-uniformity is very well dealt with by a cold plasma approach (see section 4.3) and the reason for this is clear: the essential feature is the asymmetry and this is well described by cold plasma theory. Part of chapter 2, chapters 3, 4 and 6 will show the great usefulness of the description of a bounded plasma by a *constant* equivalent permittivity ϵ_p (tensorial or scalar) and the good agreement between such theoretical description and experiments. As underlined above, one must be very cautious when using an inhomogeneous ϵ_p.

b. Variational procedures

Very general variational principles have been formulated by Sturrock[71] and Low[72] for application to small-amplitude disturbances in relativistic as well as non-relativistic multi-stream problems. A more specialized form of these principles (see equation (41) of reference 69) is

$$\delta \int_{\substack{\text{All} \\ \text{space}}} \left\{ \frac{m_e \langle N_e \rangle}{2} \left[\left(\frac{\partial}{\partial t} + \langle \mathbf{U}_e \rangle \cdot \nabla \right) \boldsymbol{\xi} \right]^2 + e \langle N_e \rangle \boldsymbol{\xi} \cdot \nabla \phi + \frac{\epsilon}{2} (\nabla \phi)^2 \right\} d^3\mathbf{r}\, dt = 0 \tag{1-40}$$

where the as yet undefined quantity is the electron displacement vector $\boldsymbol{\xi}(\mathbf{r}, t)$. If $\langle \mathbf{U}_e \rangle = 0$, the variational principle (1-40) leads to the following

Oscillations and Variational Procedures

Euler-Lagrange equations:

$$m_e \frac{\partial^2 \xi}{\partial t^2} = e \nabla \phi \qquad (1\text{-}41)$$

$$\nabla^2 \phi = -\frac{e}{\epsilon_0} \nabla \cdot (\langle N_e \rangle \xi) \qquad (1\text{-}42)$$

Taking $\xi(\mathbf{r}, t) = \xi(\mathbf{r}) \sin \omega t$ and $\phi(\mathbf{r}, t) = \phi(\mathbf{r}) \sin \omega t$, equation (1-41) yields

$$\xi(\mathbf{r}, t) = -\frac{e \nabla \phi(\mathbf{r})}{\omega^2 m_e} \sin \omega t \qquad (1\text{-}43)$$

Introducing (1-43) in (1-40) and integrating over t between 0 and π/ω, one obtains

$$\delta \int_{\text{All space}} \frac{\epsilon (\nabla \phi)^2}{2} d^3\mathbf{r} = 0 \qquad (1\text{-}44)$$

where ϵ is equal to $\epsilon_0 (1 - \omega_e^2/\omega^2)$ when $\omega_e \neq 0$, i.e. when plasma is present. The Euler-Lagrange equation deduced from (1-44) is precisely

$$\nabla \cdot (\epsilon \nabla \phi) = 0 \qquad (1\text{-}45)$$

which is valid in dielectrics or equivalent dielectric media. Using (1-45), we have

$$\epsilon (\nabla \phi)^2 = \nabla \cdot (\phi \epsilon \nabla \phi) - \phi \nabla \cdot (\epsilon \nabla \phi) = \nabla \cdot (\phi \epsilon \nabla \phi) \qquad (1\text{-}46)$$

Using (1-46) and the theorem of Gauss–Ostrogradski, we obtain

$$I = \int_{\text{All space}} \frac{\epsilon}{2} (\nabla \phi)^2 d^3\mathbf{r} = \frac{1}{2} \int_{S_\infty} \phi \epsilon (\nabla \phi \cdot \mathbf{n}) \, dS = 0 \qquad (1\text{-}47)$$

where $\mathbf{n}$ is the unit vector perpendicular to the surface element dS and where S_∞ is the surface at infinity. The integral over S_∞ vanishes if $\phi \nabla \phi$ tends sufficiently rapidly towards zero when $R \to \infty$. Equation (1-47), which is equivalent to equation (1-44) when ϕ has a correct behaviour at infinity, can be written as

$$I = \int_{\text{All space}} \frac{\epsilon}{2} (\nabla \phi)^2 d^3\mathbf{r} = \int_{\text{Plasma}} \frac{\epsilon_0}{2} (1 - \omega_e^2/\omega^2)(\nabla \phi_p)^2 d^3\mathbf{r}$$
$$+ \int_{\text{Ext}} \frac{\epsilon}{2} (\nabla \phi_{\text{ext}})^2 d^3\mathbf{r} = 0 \qquad (1\text{-}48)$$

where the ϕ's are the exact solutions in the different media. This is very general and useful form of the resonance condition for cold plasma systems.

Now, since equation (1-48) is obtained from a variational principle, we know that if we introduce a trial function ϕ_{pt} in (1-48) instead of the exact ϕ_p the error made on the eigenvalue ω_R is of the second order with respect to the errors made in the trial function ϕ_{pt}. Using equation (1-46) and the theorem of Gauss–Ostrogradski, the second term of the right-hand side of equation (1-48) becomes

$$\int_{\text{Ext}} (\nabla \phi_{\text{ext}})^2 \, d^3\mathbf{r} = \int_{S_\infty} \phi_{\text{ext}} \epsilon (\nabla \phi_{\text{ext}} \cdot \mathbf{n}) \, dS - \int_{S_p} \phi_{\text{ext}} \epsilon (\nabla \phi_{\text{ext}} \cdot \mathbf{n}) \, dS \quad (1\text{-}49)$$

where $\mathbf{n}$ is the unit vector perpendicular to the surface element dS and where S_∞ and S_p are the surface at infinity and the plasma surface respectively. Again, the integral over S_∞ is equal to zero. Substituting (1-49) in equation (1-48) where ϕ_{pt} is introduced instead of ϕ_p one obtains

$$\omega_R^2 = \frac{\int_{\text{plasma}} \epsilon_0 \omega_e^2(\mathbf{r}) (\nabla \phi_{pt})^2 \, d^3\mathbf{r}}{\int_{\text{plasma}} \epsilon_0 (\nabla \phi_{pt})^2 \, d^3\mathbf{r} - \int_{S_p} \phi_{\text{ext}} \epsilon (\nabla \phi_{\text{ext}} \cdot \mathbf{n}) \, dS} \quad (1\text{-}50)$$

where the error made on the eigenvalue ω_R is of the second order with respect to the errors made in the trial function ϕ_{pt}. This cold plasma form of the variational principle has been used by Crawford and Kino[73] to calculate cold plasma eigenfrequencies to the second order with respect to the errors made in the trial function used for ϕ_p.

Applying (1-50) to an inhomogeneous plasma column or to the inhomogeneous plasma slab-condenser system of section 2.2 and using the uniform plasma solutions as trial functions, one finds exactly the same results as for the uniform case but with ω_e replaced, of course, by $(\omega_e)_{\text{Av}}$, i.e. $(\omega_e)_{\text{Av}}/\sqrt{2}$ and $(\omega_e)_{\text{Av}}\sqrt{((l-a)/l)}$ respectively. Compare this to the exact results of the cold plasma cylinder described by figure 1.1.

One must remember, however, the severe restrictions resulting from the analysis of the properties of an *inhomogeneous* cold plasma (section 1.5a). These restrictions are such as to considerably reduce the practical use of formula 1.50)* since the previous analysis has shown that the results of the equivalent (in the sense $\omega_e = \omega_{e\text{Av}}$) uniform cold plasma are in most instances (not including the asymmetrical inhomogeneity effects) more reliable than that of the non-uniform one.

---

* The basic properties of the model result in a conspicuous lack of convergence when the variational procedure is applied to a plasma with a spatially bounded continuous charge density[70].

1.6 Moments method

As is well known (see for example reference 2 or 12), it is possible to deduce the macroscopic hydrodynamic equations of a plasma by taking successive velocity moments of the Boltzmann equation which has, in Cartesian coordinates, the following form

$$\frac{\partial F}{\partial t} + \mathbf{v} \cdot \frac{\partial F}{\partial \mathbf{r}} + \frac{d\mathbf{v}}{dt} \cdot \frac{\partial F}{\partial \mathbf{v}} = \left(\frac{\partial F}{\partial t}\right)_{\text{coll}} \quad (1\text{-}51)$$

where $F(\mathbf{r}, \mathbf{v}, t)$ is the distribution function, $\mathbf{r}$ is the position vector, and $\mathbf{v}$ the velocity. One must then operate a judicious cut in the infinite chain of moments of this equation in order to retain a limited number of macroscopic equations, and this truncation must be done with least violence to the physics. Some authors neglect a certain moment (e.g. the heat flow) while others neglect a certain moment equation and postulate an equation of state involving the highest moment retained.

In 1960 Obermann[74,75] and, independently, Vandenplas and Gould[60,21,61] showed that the use of a finite set of moment equations was adequate to describe the plasma when the width of the distribution was in some sense small. This description corresponds in fact to a systematic expansion of the distribution function F in powers of the temperature T. The latter authors have also shown that it is useless to use too many moment equations when the linearization procedure already involves neglecting terms of a certain magnitude. Obermann's approach, which is based on the use of a delta function and its derivatives in velocity space (small thermal spread), is mathematically equivalent to the approach by Vandenplas and Gould. We will follow the latter treatment.

a. Dispersion relation

Consider the collisionless Boltzmann (or Boltzmann–Vlasov) equation which gives quite a good description of low density plasmas. The criterion for using a collision-free plasma is in fact $\nu/\omega \ll 1$ which, for our purposes, corresponds to $\nu/\omega_e \ll 1$. We analyse plane longitudinal waves of the form $e^{i(kx-\omega t)}$ in an infinite plasma with uniform equilibrium density. The linearized Vlasov equation is then

$$-i\omega f + ikvf - \frac{e}{m_e} E \frac{\partial \langle F \rangle}{\partial v} = 0 \quad (1\text{-}52)$$

with $F = \langle F \rangle + f$ ($\langle F \rangle$ and f being respectively the equilibrium and perturbed distribution functions) and $v \equiv v_x$. We define the total density

N, the perturbed velocity u, the pressure P and the moment of order j, M_j, through

$$N = \langle N \rangle + n = \int F\, dv \tag{1-53}$$

$$Nu = \int vF\, dv \tag{1-54}$$

$$\frac{P}{m_e} = \frac{\langle P \rangle}{m_e} + \frac{p}{m_e} = \int w^2 F\, dv \quad \text{with} \quad w = v - u \tag{1-55}$$

$$\frac{M_j}{m_e} = \frac{\langle M_j \rangle}{m_e} + \frac{m_j}{m_e} = \int w^j F\, dv \tag{1-56}$$

where the integration over v_y and v_z has already been performed and is not mentioned. The brackets indicate the equilibrium quantities and the small letters indicate the perturbed quantities. We take the successive moments of equation (1-52) with respect to v and, after assuming that $\langle F \rangle$ is even and centered on zero, we obtain

$$\omega n - k\langle N \rangle u = 0 \tag{1-57}$$

This is the continuity equation.

$$i\omega \langle N \rangle u - \frac{ikp}{m_e} - \frac{e\langle N \rangle E}{m_e} = 0 \tag{1-58}$$

Equation (1-58) is the equation of momentum transfer.

$$i\omega p - ik(3u\langle P \rangle + m_3) = 0 \tag{1-59}$$

$$i\omega(3u\langle P \rangle + m_3) - ikm_4 - 3e\frac{\langle P \rangle}{m_e}E = 0 \tag{1-60}$$

$$i\omega m_4 - ik(5u\langle M_4 \rangle + m_5) = 0 \tag{1-61}$$

$$i\omega(5u\langle M_4 \rangle + m_5) - ikm_6 - 5e\frac{\langle M_4 \rangle}{m_e}E = 0 \tag{1-62}$$

etc. ...

Setting successively $p = 0$, $m_3 = 0$, $m_4 = 0$, $m_5 = 0$, $m_6 = 0, \ldots,$ $m_{2r+2} = 0, \ldots,$ eliminating m_j between the last equation where it appears and the preceding ones, and taking into account that div $\mathbf{E} = ikE = -en/\epsilon_0$, we obtain

$$\omega^2 = \omega_e^2 \tag{1-63}$$

$$\omega^2 = \omega_e^2 \frac{1}{1 - \dfrac{3k^2\langle P \rangle}{m_e \omega^2 \langle N \rangle}} \tag{1-64}$$

$$\omega^2 = \omega_e^2\left(1 + \frac{3k^2\langle P \rangle}{m_e\omega^2\langle N \rangle}\right) = \omega_e^2\left(1 + \frac{3k^2 v_0^2}{\omega^2}\right) \tag{1-65}$$

Moments Method

This is the well-known Bohm–Gross form of the dispersion relation[90,91]

$$\omega^2 = \omega_e^2 \frac{1 + \dfrac{3k^2\langle P\rangle}{m_e\omega^2\langle N\rangle}}{1 - \dfrac{5k^4\langle M_4\rangle}{m_e\omega^2\langle N\rangle}} \tag{1-66}$$

. . . .

$$\omega^2 = \omega_e^2\left[1 + \sum_{r=1}^{l}(2r+1)!!\left(\sqrt{\frac{KT}{m_e}}\frac{k}{\omega}\right)^{2r}\right], \quad l \to \infty \tag{1-67}$$

This last dispersion relation is the same as that obtained for the real part of ω by Landau's solution[57] of the Vlasov equation when $KTk^2/m_e\omega^2 \ll 1$. It is an asymptotic series.

b. Approximations made in the moments method

We first seek an expression giving f as a function of v. Using equation (1-52), div $\mathbf{E} = -en/\epsilon_0$ with $\mathbf{E} = -\operatorname{grad} \phi$ (quasi-static approximation if the waves are not purely longitudinal in the sense curl $\mathbf{H} \equiv 0$; for more details see section 2.1) and equation (1-57), we obtain

$$f = \frac{\omega_e^2}{\omega^2}\frac{u}{\left(\dfrac{kv}{\omega} - 1\right)}\frac{\partial\langle F\rangle}{\partial v} = \frac{m_e}{KT}\frac{\omega_e^2}{\omega^2}\frac{uv}{\left(1 - \dfrac{kv}{\omega}\right)}\langle F\rangle \tag{1-68}$$

if $\langle F\rangle$ is Maxwellian. Equation (1-68) clearly shows that something peculiar will happen when $v = \omega/k$ (that is, for the particles having a velocity equal to the phase velocity of the wave) and that the linearization assumption is very questionable for these velocities.

Equation (1-68) can be used to assess the order of magnitude of the term neglected in the linearization of the Vlasov equation. The $\partial f/\partial v$ term was neglected with respect to the $\partial\langle F\rangle/\partial v$ term and it is readily seen that

$$\left|\frac{\partial f/\partial v}{\partial\langle F\rangle/\partial v}\right| \ll 1 \quad \text{is equivalent to} \quad \left|\frac{ku}{\omega}\right| \ll 1$$

except when $v \simeq \omega/k$ and when $v \simeq 0$ (then $\partial\langle F\rangle/\partial v \simeq 0$). One therefore neglects terms of order $|ku/\omega|$ with respect to those retained, when one derives the hydrodynamic equations from the linearized Vlasov equation.

We now assume that $\langle F\rangle$ is characterized, in fact, by a maximum velocity $v_{\max}$ (remember that the velocity of light c is, in any case, a maximum velocity). Thus for k sufficiently small, we have $|kv/\omega| < 1$ for all $v < v_{\max}$ such that $\langle F \rangle \neq 0$. Van Kampen[76] has shown that if a maximum value of the velocity $v_{\max}$ does not exist, but if $\langle F \rangle \simeq 0$ for this value, the difficulties are essentially of mathematical nature but that the results (dispersion relation, etc.) remain valid with the difference that the oscillations resulting from an initial perturbation no longer exist during an infinite length of time but instead during a finite time. We therefore choose a 'truncated' Maxwellian velocity distribution for $\langle F \rangle$ such that $\langle F \rangle = 0$ for $v > v_{\max}$ with $v_{\max} = 3v_0$, ($v_0 = \sqrt{(KT/m)}$ is the root mean square velocity in the x-direction); then

$$\left| \frac{kv_{\max}}{\omega} \right| = \left| \frac{3kv_0}{\omega} \right| < 1 \quad \text{or} \quad \left| \frac{kv_0}{\omega} \right|^2 \ll 1 \qquad (1\text{-}69)$$

It is in this last sense that the thermal spread must be small. Equation (1-68) yields then rigorously

$$f = \frac{m_e}{KT} \frac{\omega_e^2}{\omega^2} uv \langle F \rangle \left[1 + \frac{kv}{\omega} + \left(\frac{kv}{\omega}\right)^2 + \left(\frac{kv}{\omega}\right)^3 + \cdots \right] \qquad (1\text{-}70)$$

If we introduce (1-70) in the definitions (1-53)–(1-56) of the moments and if we use the dispersion relation (1-67), we find the following linearized expressions

$$n = \langle N \rangle \frac{ku}{\omega} + 0 \left(\frac{kv_0}{\omega}\right)^4 \qquad (1\text{-}71)$$

$$\frac{p}{m_e} = 3 \langle N \rangle v_0^2 \frac{ku}{\omega} + 6 \langle N \rangle v_0^2 \frac{ku}{\omega} \left(\frac{kv_0}{\omega}\right)^2 + \cdots \qquad (1\text{-}72)$$

$$\frac{m_3}{m_e} = 6 \langle N \rangle v_0^2 u \left(\frac{kv_0}{\omega}\right)^2 + \cdots \qquad (1\text{-}73)$$

$$\frac{m_4}{m_e} = 15 \langle N \rangle v_0^4 \frac{ku}{\omega} + 60 \langle N \rangle v_0^4 \frac{ku}{\omega} \left(\frac{kv_0}{\omega}\right)^2 + \cdots \qquad (1\text{-}74)$$

These expressions of the moments p, m_3, $m_4 \cdots$ allow us to examine the approximation made when p is neglected in equation (1-58), m_3 in equation (1-59), m_4 in equation (1-60), etc.

The perturbed pressure p can be neglected in (1-58) if

$$\left| \frac{kp/m_e}{\omega \langle N \rangle u} \right| \ll 1 \quad \text{and} \quad \left| \frac{kp}{e \langle N \rangle E} \right| \ll 1 \qquad (1\text{-}75)$$

Using (1-72), one sees that conditions (1-75) are satisfied if

$$\left| \frac{ku}{\omega} \right| \ll 1 \quad \text{and} \quad \left| \frac{kv_0}{\omega} \right|^2 \ll 1 \qquad (1\text{-}76)$$

Moments Method

and the procedure is thus consistent if $|ku/\omega| \sim |kv_0/\omega|^2 \ll 1$; this leads to the cold linearized plasma theory. Using (1-73) one sees that m_3 can be neglected in equation (1-59) if $|ku/\omega| \ll 1$, $|kv_0/\omega|^2 \ll 1$, and $|u/v_0| \ll 1$, provided

$$\left|\frac{ku}{\omega}\right| \leqslant \left|\frac{kv_0}{\omega}\right|^2 \tag{1-77}$$

Similarly, m_4 can be neglected in equation (1-60) and m_5 in equation (1-61) if the quantities $|ku/\omega|$, $|kv_0/\omega|^2$, $|u/v_0|^2$ multiplied by integers are much smaller than 1.

We are now in a position to assess the approximations which are made when the chain of hydrodynamic equations is broken off by setting a certain perturbed moment m_i equal to zero. We successively neglect p in (1-58), m_3 in (1-59), m_4 in (1-60), etc. Each time we carry out upwards the successive substitutions and introduce the value found for p in the momentum transfer equation (1-58) which we rewrite

$$i\omega m_e \langle N \rangle u - e\langle N \rangle E - ikp + 0(ku/\omega)^2 = 0 \tag{1-78}$$

remembering that in the linearization process we have neglected the terms which are $|ku/\omega|$ smaller than those retained.

When p is set equal to zero in (1-78), we have, as already stated, the cold linearized description of plasma oscillations. When m_3 is set equal to zero in (1-59), we obtain

$$p = 3\langle N \rangle m_e v_0^2 \frac{ku}{\omega} \tag{1-79}$$

When m_4 is set equal to zero in (1-60), we obtain after substitution

$$m_3 = -9\langle N \rangle m_e v_0^2 u \left(\frac{kv_0}{\omega}\right)^2 \tag{1-80}$$

$$p = 3\langle N \rangle m_e v_0^2 \frac{ku}{\omega}\left[1 - 3\left(\frac{kv_0}{\omega}\right)^2\right] \tag{1-81}$$

When m_5 is set equal to zero in (1-61), we obtain after successive substitutions

$$m_4 = 5\langle N \rangle m_e v_0^4 \frac{ku}{\omega} \tag{1-82}$$

$$m_3 = 6\langle N \rangle m_e v_0^2 u \left(\frac{kv_0}{\omega}\right)^2 \tag{1-83}$$

$$p = 3\langle N \rangle m_e v_0^2 \frac{ku}{\omega}\left[1 + 2\left(\frac{kv_0}{\omega}\right)^2\right] \tag{1-84}$$

When m_6 is set equal to zero in (1-62), we obtain

$$m_5 = -45\langle N\rangle m_e v_0^4 u \left(\frac{kv_0}{\omega}\right)^2 \tag{1-85}$$

$$m_4 = 15\langle N\rangle m_e v_0^4 \frac{ku}{\omega}\left[1 - 3\left(\frac{kv_0}{\omega}\right)^2\right] \tag{1-86}$$

$$m_3 = 6\langle N\rangle m_e v_0^2 u \left(\frac{kv_0}{\omega}\right)^2\left[1 - \frac{15}{2}\left(\frac{kv_0}{\omega}\right)^2\right] \tag{1-87}$$

$$p = 3\langle N\rangle m_e v_0^2 \frac{ku}{\omega}\left[1 + 2\left(\frac{kv_0}{\omega}\right)^2 - 15\left(\frac{kv_0}{\omega}\right)^4\right] \tag{1-88}$$

The meaning of the truncation scheme becomes very clear when the successive values (1-79), (1-81), (1-84), (1-88) found for p are introduced in the momentum transfer equation (1-78). Neglecting m_3 and working upwards introduces the correct expression of p to first order in T [i.e. in $kv_0/\omega)^2$] in equation (1-78). Neglecting m_5 and working upwards introduces the correct expression of p to second order in T [i.e. in $(kv_0/\omega)^4$] in equation (1-78). Generalizing, one finds that the neglect of m_{2r+1} in the $(2r + 1)$-th hydrodynamic equation and the subsequent substitutions introduce the correct expression of p to r-th order in T [i.e. in $(kv_0/\omega)^{2r}$] in equation (1-78).

Now, since the linearization procedure involved the neglect of terms of order ku/ω with respect to those retained, there is no use in retaining in the expression of p, powers of kv_0/ω which are such that $|kv_0/\omega|^{2r} < |ku/\omega|$ as the precision gained would be illusory. In fact, when

$$|ku/\omega| \sim |kv_0/\omega|^{2r} \tag{1-89}$$

m_{2r+1} must be neglected and it is illusory to take more than $(2r + 1)$ hydrodynamic equations into account to give a coherent description of the linearized plasma oscillations.

The reader will note that the first four hydrodynamic equations (1-57)–(1-60) give all the information contained in the Vlasov equation to the first order in the temperature T. The fifth hydrodynamic equation (1-61) contains only information in T^2 or higher as it involves moments of order 4 or more, etc. This fact accounts for the stability of the various terms appearing in the dispersion relation, in the expression for p, etc. Once all the information in a given power of the temperature (e.g. in T^1) which is contained in the Vlasov equation has been used, the term appearing to that order in the dispersion relation, in p, in m_3, ..., etc. is not altered by the use of further hydrodynamic equations higher powers of the temperature.

In the one-dimensional case treated above, it was nevertheless possible to obtain the correct dispersion relation and the correct equation of state for p to first order in T by using only the first three hydrodynamic equations. When a steady magnetic field is present, this is no longer possible and a full description to order T entails the use of the first four moments (0, 1, 2 and 3) of the Vlasov equation, see reference 77 and section 5.2g.

Using the dispersion relation (1-67), one immediately sees that the above approach is only strictly valid in the region where ω is rather close to ω_e.

1.7 Boltzmann—Vlasov equation in cylindrical coordinates

The expression (1-51) of the Boltzmann–Vlasov equation is only valid in generalized Cartesian coordinates. When working in another coordinate system (e.g. in cylindrical coordinates), it is often useful to be able to write down this equation in such a coordinate system. Following Vandenplas and Gould[21,60,61], we give two general methods to do this conveniently, but treat the special case of cylindrical coordinates.

We wish to write down the collisionless Boltzmann (or Vlasov) equation for the distribution function F in joint cylindrical position (r, θ, z) and velocity (v_r, v_θ, v_z) space.

First derivation. We start from the six-dimensional phase space $(q_1, q_2, q_3, p_1, p_2, p_3)$ with $q_1 = r, q_2 = \theta, q_3 = z, p_1 = p_r, p_2 = p_\theta, p_3 = p_z$. This phase space has a volume element $dq_1\, dq_2\, dq_3\, dp_1\, dp_2\, dp_3$ which has a Cartesian form. In such a space, the expression of the divergence (and thus the continuity equation) is the same as in a six-dimensional space referred to a Cartesian coordinate system. When electromagnetic forces are included, one has the following expressions for the generalized momenta

$$p_r = m\dot{r} + qA_r; \quad p_\theta = mr^2\dot{\theta} + qrA_\theta; \quad p_z = m\dot{z} + qA_z \quad (1\text{-}90)$$

where q is the electric charge, $\mathbf{A}$ is the potential vector, and the dot indicates derivation with respect to time t. One has the Hamiltonian

$$H = \frac{1}{2m}\left[(p_r - qA_r)^2 + \frac{(p_\theta - qrA_\theta)^2}{r^2} + (p_z - qA_z)^2\right] \quad (1\text{-}91)$$

The phase space $(r, \theta, z, p_r, p_\theta, p_z)$ is a generalized Cartesian coordinate system where the number of particles dN contained in a volume element $dr\, d\theta\, dz\, dp_r\, dp_\theta\, dp_z$ is given through the distribution function F_p by

$$dN = F_p\, dr\, d\theta\, dz\, dp_r\, dp_\theta\, dp_z \quad (1\text{-}92)$$

The continuity equation in the six-dimensional phase space, with no collisions, gives

$$\frac{\partial F_p}{\partial t} + \sum_{i=1}^{3} \frac{\partial}{\partial q_i}(\dot{q}_i F_p) + \sum_{i=1}^{3} \frac{\partial}{\partial p_i}(\dot{p}_i F_p) = 0. \tag{1-93}$$

But using the Hamiltonian equations of motion, one has

$$\frac{\partial \dot{q}_i}{\partial q_i} + \frac{\partial \dot{p}_i}{\partial p_i} = 0. \tag{1-94}$$

Thus

$$\frac{\partial F_p}{\partial t} + \dot{r}\frac{\partial F_p}{\partial r} + \dot{\theta}\frac{\partial F_p}{\partial \theta} + \dot{z}\frac{\partial F_p}{\partial z}$$
$$+ \left[m\ddot{r} + q\left(\dot{r}\frac{\partial A_r}{\partial r} + \dot{\theta}\frac{\partial A_r}{\partial \theta} + \dot{z}\frac{\partial A_r}{\partial z} + \frac{\partial A_r}{\partial t}\right)\right]\frac{\partial F_p}{\partial p_r}$$
$$+ \left[m(2\dot{r}\dot{\theta} + r^2\ddot{\theta}) \right.$$
$$\left. + q\left(\dot{r}A_\theta + r\dot{r}\frac{\partial A_\theta}{\partial r} + r\dot{\theta}\frac{\partial A_\theta}{\partial \theta} + r\dot{z}\frac{\partial A_\theta}{\partial z} + r\frac{\partial A_\theta}{\partial t}\right)\right]\frac{\partial F_p}{\partial p_\theta}$$
$$+ \left[m\ddot{z} + q\left(\dot{r}\frac{\partial A_z}{\partial r} + \dot{\theta}\frac{\partial A_z}{\partial \theta} + \dot{z}\frac{\partial A_z}{\partial z} + \frac{\partial A_z}{\partial t}\right)\right]\frac{\partial F_p}{\partial p_z} = 0 \tag{1-95}$$

Let us change to coordinates $r = r$, $\theta = \theta$, $z = z$, $v_r = (p_r - qA_r)/m$, $v_\theta = (p_\theta - qrA_\theta)/mr$, $v_z = (p_z - qA_z)/m$, $t = t$ and note $F_p{}'$ the former F_p when it is expressed as a function of these new variables. After a somewhat lengthy calculation we observe that all the terms involving the components of **A** cancel out and we finally obtain

$$\frac{\partial F_p{}'}{\partial t} + v_r\left[\frac{\partial F_p{}'}{\partial r} + \frac{\partial F_p{}'}{\partial v_\theta}\left(\frac{-v_\theta}{r}\right)\right] + \frac{v_\theta}{r}\frac{\partial F_p{}'}{\partial \theta} + v_z\frac{\partial F_p{}'}{\partial z}$$
$$+ \ddot{r}\frac{\partial F_p{}'}{\partial v_r} + (2\dot{r}\dot{\theta} + r\ddot{\theta})\frac{\partial F_p{}'}{\partial v_\theta} + \ddot{z}\frac{\partial F_p{}'}{\partial v_z} = 0. \tag{1-96}$$

Now we go from the generalized Cartesian system $r, \theta, z, p_r, p_\theta, p_z$ over to the curved (or physical) cylindrical system $r, \theta, z, v_r, v_\theta, v_z$ in which the distribution function F is defined such that

$$dN = F_p{}'(r, \theta, z, v_r, v_\theta, v_z, t)\, dr\, d\theta\, dz\, m^3\, r\, dv_r\, dv_\theta\, dv_z$$
$$= F(r, \theta, z, v_r, v_\theta, v_z, t)r\, dr\, d\theta\, dz\, dv_r\, dv_\theta\, dv_z. \tag{1-97}$$

whence

$$m^3 F_p{}' = F. \tag{1-98}$$

Using

$$\phi_r = \ddot{r} - r\dot{\theta}^2; \qquad \phi_\theta = r\ddot{\theta} + 2\dot{r}\dot{\theta}; \qquad \phi_z = \ddot{z} \tag{1-99}$$

Description of a Warm Non-Uniform Plasma

where ϕ_r, ϕ_θ and ϕ_z are the components of the acceleration, we obtain

$$\frac{\partial F}{\partial t} + v_r \frac{\partial F}{\partial r} + \frac{v_\theta}{r}\frac{\partial F}{\partial \theta} + v_z \frac{\partial F}{\partial z} + \phi_r \frac{\partial F}{\partial v_r} + \phi_\theta \frac{\partial F}{\partial v_\theta}$$

$$+ \phi_z \frac{\partial F}{\partial v_z} + \frac{v_\theta^2}{r}\frac{\partial F}{\partial v_r} - \frac{v_\theta v_r}{r}\frac{\partial F}{\partial v_\theta} = 0 \quad (1\text{-}100)$$

which is the Boltzmann–Vlasov equation in cylindrical coordinates.

Second derivation. Equation (1-100) can be established more rapidly by immediately writing down the continuity equation for a distribution function F_V in the generalized Cartesian system $r, \theta, z, v_r, v_\theta, v_z$ with a Cartesian volume element $dr\, d\theta\, dz\, dv_r\, dv_\theta\, dv_z$

$$\frac{\partial F_V}{\partial t} + \frac{\partial}{\partial r}(\dot{r} F_V) + \frac{\partial}{\partial \theta}(\dot{\theta} F_V) + \frac{\partial}{\partial z}(\dot{z} F_V) + \frac{\partial}{\partial v_r}(\dot{v}_r F_V)$$

$$+ \frac{\partial}{\partial v_\theta}(\dot{v}_\theta F_V) + \frac{\partial}{\partial v_z}(\dot{v}_z F_V) = 0 \quad (1\text{-}101)$$

In this way the first and lengthy step involving the corresponding phase space is avoided. Using equation (1-99), we have

$$\frac{\partial F_V}{\partial t} + v_r \frac{\partial F_V}{\partial r} + \frac{v_\theta}{r}\frac{\partial F_V}{\partial \theta} + v_z \frac{\partial F_V}{\partial z} + \phi_r \frac{\partial F_V}{\partial v_r} + \phi_\theta \frac{\partial F_V}{\partial v_\theta}$$

$$+ \frac{v_\theta^2}{r}\frac{\partial F_V}{\partial v_r} - \frac{v_r v_\theta}{r}\frac{\partial F_V}{\partial v_\theta} + \phi_z \frac{\partial F_V}{\partial v_z} - \frac{v_r}{r} F_V = 0. \quad (1\text{-}102)$$

To obtain equation (1-102) we made the assumption that $\partial \phi_i/\partial v_i = 0$. This assumption is valid for electromagnetic forces and is the same as the one resulting from the form taken for (1-90) and (1-91).

We go over from the generalized Cartesian system $r, \theta, z, v_r, v_\theta, v_z$ to the curved (or physical) system $r, \theta, z, v_r, v_\theta, v_z$ through

$$dN = F_V\, dr\, d\theta\, dz\, dv_r\, dv_\theta\, dv_z = F r\, dr\, d\theta\, dz\, dv_r\, dv_\theta\, dv_z. \quad (1\text{-}103)$$

whence

$$F_V = rF \quad (1\text{-}104)$$

and we again obtain equation (1-100).

The two methods described above are quite general and enable one to write down the Boltzmann–Vlasov equation in any coordinate system.

1.8 Description of a warm non-uniform plasma with tensorial perturbed pressure

If we extend the conclusions of section 1.7 to a bounded inhomogeneous three-dimensional plasma, and if we wish to describe a plasma so as to include terms of order $(kv_0/\omega)^2$ in the pressure, i.e. neglect $|ku/\omega| \sim |kv_0/\omega|^4$, we must take the moments of order zero, one and two of the Boltzmann–Vlasov equation and set m_3 equal to zero. In this way we obtain a correct equation of state to order T for the perturbed tensorial pressure. A still more elaborate way of proceeding would be to include the moments of order three of the Vlasov equation and set m_4 equal to zero in view of using all the information to order T provided by the Vlasov equation.

a. Cylindrical case

The following model is adopted for a cylindrical plasma column in equilibrium. The electron density is $\langle N \rangle = \langle N_0 \rangle g(r)$ where $\langle N_0 \rangle$ is the equilibrium density on the axis of the column. The density gradient is balanced by a static electric field strength $\langle E \rangle$ resulting from the non-neutralization of positive and negative electric charges. We assume that the electron drift velocity toward the wall is zero so that the distribution function $\langle F \rangle$ is Maxwellian and centred on zero. The small ambipolar drift is therefore neglected (see Thompson reference 6, chapter 3).

To write down the hydrodynamic equations in cylindrical coordinates, we take the successive moments of the Boltzmann–Vlasov equation as expressed in this system of coordinates by equation (1-100). The momentum transfer equation, obtained by taking the first-order moment of equation (1-100), then gives for the equilibrium quantities

$$\langle E_r \rangle = -\frac{KT}{eg(r)} \frac{dg(r)}{dr}. \tag{1-105}$$

The infinite plasma cylinder has a translational symmetry along its z-axis and $\partial/\partial z = 0$. We Fourier analyse in θ the perturbations E_r, E_θ, n, u_θ, u_r, p_{rr}, $p_{\theta\theta}$, $p_{r\theta}$ and adopt the notations:

$$\frac{P_{ij}}{m_e} = \frac{\langle P_{ij} \rangle}{m_e} + \frac{p_{ij}}{m_e} = \int_{-\infty}^{+\infty} w_i w_j F \, dv^3 \quad \text{with} \quad w_i = v_i - u_i.$$

The fact that $\partial/\partial z = 0$ means that there is no perturbation along the z-axis and that the perturbed distribution function is of the form $f = f(v_r, v_\theta)\delta(v_z)$ where δ is the delta function.

$$E_\theta = \sum_{m=1}^{\infty} E_{\theta m}(r) \cos m\theta; \qquad E_r = \sum_{m=1}^{\infty} E_{rm}(r) \sin m\theta$$

$$n = \sum_{m=1}^{\infty} n_m(r) \sin m\theta; \qquad u_\theta = \sum_{m=1}^{\infty} u_{\theta m}(r) \cos m\theta$$

$$u_r = \sum_{m=1}^{\infty} u_{rm}(r) \sin m\theta; \qquad p_{rr} = \sum_{m=1}^{\infty} p_{rrm}(r) \sin m\theta$$

$$p_{\theta\theta} = \sum_{m=1}^{\infty} p_{\theta\theta m}(r) \sin m\theta; \qquad p_{\theta r} = p_{r\theta} = \sum_{m=1}^{\infty} p_{\theta r m}(r) \cos m\theta$$

Description of a Warm Non-Uniform Plasma

Considering again perturbations with an $e^{-i\omega t}$ time-dependence, we have the following equation for the zero-th moment of the Boltzmann equation (continuity equation):

$$-i\omega n_m + u_{rm}\langle N_0\rangle \frac{dg}{dr} + \langle N_0\rangle g\left[\frac{u_{rm}}{r} + \frac{du_{rm}}{dr} - \frac{m}{r}u_{\theta m}\right] = 0 \quad (1\text{-}106)$$

The moment of order one gives the two momentum transfer equations:

$$-i\omega m_e \langle N_0\rangle g u_{rm} + \frac{dp_{rrm}}{dr} - \frac{m}{r}p_{r\theta m} + \frac{p_{rrm}}{r} - \frac{p_{\theta\theta m}}{r}$$

$$-\frac{KT}{g}\frac{dg}{dr}n + eE_{rm}\langle N_0\rangle g = 0 \quad (1\text{-}107)$$

$$-i\omega m_e \langle N_0\rangle g u_{\theta m} + \frac{dp_{\theta rm}}{dr} + \frac{m}{r}p_{\theta\theta m} + \frac{2p_{\theta rm}}{r} + eE_{\theta m}\langle N_0\rangle g = 0 \quad (1\text{-}108)$$

The moment of order two gives three transport equations for the tensorial pressure after having neglected the third-order moments:

$$p_{rrm} = \frac{\langle N_0\rangle KT}{i\omega}\left[g\left(3\frac{du_{rm}}{dr} - \frac{m}{r}u_{\theta m} + \frac{u_{rm}}{r}\right) + \frac{dg}{dr}u_{rm}\right] \quad (1\text{-}109)$$

$$p_{\theta\theta m} = \frac{\langle N_0\rangle KT}{i\omega}\left[g\left(\frac{du_{rm}}{dr} - \frac{3m}{r}u_{\theta m} + \frac{3u_{rm}}{r}\right) + \frac{dg}{dr}u_{rm}\right] \quad (1\text{-}110)$$

$$p_{r\theta m} = \frac{\langle N_0\rangle KT}{i\omega}g\left[\frac{du_{\theta m}}{dr} + \frac{m}{r}u_{rm} - \frac{u_{\theta m}}{r}\right] \quad (1\text{-}111)$$

The Maxwell equations for the divergence and the curl of **E** with the quasi-static approximation for the latter, give

$$\frac{E_{rm}}{r} + \frac{dE_{rm}}{dr} - \frac{m}{r}E_{\theta m} = \frac{-ne}{\epsilon_0} \quad (1\text{-}112)$$

$$\frac{E_{\theta m}}{r} + \frac{dE_{\theta m}}{dr} - \frac{m}{r}E_{rm} = 0 \quad (1\text{-}113)$$

This system of eight equations with eight unknowns can be reduced to a sixth-order normal linear differential system with variable coefficients where the six unknowns are: $E_{\theta m}, E_{rm}, u_{\theta m}, u_{rm}, P_{rrm}, P_{\theta rm}$ and to two algebraical equations. This system, further simplified by the assumption of a scalar perturbed pressure $p = nKT$ which reduces the differential system to a fourth order one, is used in section 5.2 to explain the details of the secondary or temperature spectrum of electromagnetic waves scattered off a hot non-uniform plasma column. Similar results are obtained when the full system is used without making the assumption of a scalar perturbed pressure.

b. Tensorial generalization

It is interesting to write down the tensorial generalization of the equations that have been derived above[21,61]. To do this we write down, in Cartesian coordinates, the linearized equations of a plasma where the spatial inhomogeneity is balanced, as above, by a static electric field. This is perfectly straightforward and these equations can then be generalized in a unique way to obtain a tensorial form. One derives successively the continuity equation, the momentum equation and the transport equation for the tensorial pressure. Adopting Einstein's summation convention on indices that appear twice, these equations have the following form.

$$-i\omega n + \mathcal{U}^i \langle N \rangle_{,i} + \langle N \rangle \mathcal{U}^i_{,i} = 0 \tag{1-114}$$

$$m_e \langle N \rangle \frac{\partial \mathcal{U}^k}{\partial t} + \mathcal{P}^{ki}_{,i} + eN\mathcal{E}^k = 0 \tag{1-115}$$

$$\frac{\partial \mathcal{P}^{jk}}{\partial t} + \mathcal{U}^i_{,i} \langle \mathcal{P}^{jk} \rangle + \mathcal{U}^i \langle \mathcal{P}^{jk}_{,i} \rangle + \mathcal{U}^k_{,i} \langle \mathcal{P}^{ij} \rangle + \mathcal{U}^j_{,i} \langle \mathcal{P}^{ki} \rangle = 0 \tag{1-116}$$

$\mathcal{P}^{jk}$ and $\mathcal{U}^k$ are the contravariant tensors associated with the projection (or 'physical') components P_{jk} and $u_k (\mathrm{d}s^2 = g_{ij}\,\mathrm{d}q_i\,\mathrm{d}q_j)$ through

$$P_{jk} = \sqrt{g_{jj}} \sqrt{g_{kk}}\, \mathcal{P}^{jk}; \qquad u_k = \sqrt{g_{kk}}\, \mathcal{U}^k \quad \text{(No summation convention)} \tag{1-117}$$

The symbol $\mathcal{P}^{ij}_{,k}$ indicates the covariant derivative of $\mathcal{P}^{ij}$ with respect to the coordinate q_k. $\mathcal{E}^k = \langle E^k \rangle + E^k$ is the total electric field.

Problem: Use equations (1-114), (1-115) and (1-116) in cylindrical coordinates to rederive formulas (1-106)–(1-111). This is in fact a useful and independent check of the rather lengthy algebraical calculations made to obtain them.

1.9 Boundary conditions

The boundary conditions play necessarily an important role in the correct description of bounded plasmas. There is, however, in the literature some inconsistency in the use of these boundary conditions. It must be remembered that the very form of the equations used impose certain mathematical boundary conditions on the physical variables used. Now, once these mathematical matching conditions between the plasma and the outer medium are satisfied, this does not necessarily mean that the solution

Boundary Conditions

is yet determined. If the description of the plasma is rather sophisticated (e.g. a hot non-uniform plasma), one must make further physically plausible assumptions on the value of certain plasma quantities at the interface between plasma and exterior medium and these assumptions must, of course, be compatible with the mathematical conditions.

The reader will find in Appendix 1 a list of useful volume theorems together with their expression at the interface between two different media. The surface expressions of these theorems are the matching conditions that are mathematically imposed by the structure of the equations. We now briefly examine the boundary conditions for cold and hot plasmas.

a. Cold plasmas

We have seen in section 1.2 that a cold plasma, uniform or non-uniform, can be described as an equivalent dielectric medium obeying Maxwell's equations (1-23)–(1-26). The boundary conditions at the interface between the plasma and a dielectric (or a conductor) are thus the classic conditions of electromagnetic theory since the plasma behaves, in this approximation, simply as a polarizable medium with a strongly frequency-dependent tensorial permittivity (the polarization charge density is therefore, obviously, frequency-dependent).

b. Hot plasmas

The description of a hot plasma without a steady magnetic field involves at least the introduction of a perturbed scalar pressure. At the interface between the plasma and a surrounding dielectric (or conductor), one has the electromagnetic boundary conditions: continuity of the tangential component of the electric field strength and continuity of the normal component of the total current (conduction and displacement current). As any plasma pressure can be sustained by the wall, the momentum transfer equation will not furnish an additional boundary condition. One must make a physically plausible assumption. When a plasma is in contact with a solid surface, a sheath with a thickness of the order of several Debye radii is generated. Most of the electrons are reflected by the sheath back into the plasma. If the wall is a dielectric, it is coated with negative charge which repels the electrons (see Thompson reference 6). A reasonable assumption is thus that the wall is perfectly reflecting*: the perturbed distribution function f at the wall is then characterized by $f(-v) = f(v)$ (where v is the velocity perpendicular to the wall) and the odd perturbed

---

* In fact a wall (in kinetic theory of neutral particles or here) can never be perfectly reflecting since it is not flat on the molecular scale. What is meant by this expression is that electrons are scattered in such a way that $u = \int_{-\infty}^{+\infty} f(v) \, dv = 0$.

moments are zero at the wall. When a scalar pressure is used one then has the additional boundary condition that $u = 0$ (u is the mean perturbed velocity perpendicular to the wall). When a tensorial perturbed pressure is used[21,61], further boundary conditions are needed. The off-diagonal terms of the pressure tensor are odd moments of v and, using the 'specular' reflexion hypothesis consistently, they are also set equal to zero at the wall. If the wall is metallic, it is equipotential and a conduction towards it is possible; a sheath still exists and the properties of the sheath depend on the potential of the wall with respect to the potential of the plasma outside the sheath. In some instances, it seems therefore that the assumption of a practically zero perturbed conduction current at a metallic wall might also be quite satisfactory. It should however be mentioned that this sheath region close to a metal surface plays an essential role in the resonance properties of plasma systems having such a metal surface (see chapter 6).

When the equilibrium density of a plasma undergoes a strong variation over a small distance, such a variation can be approximated, for some purposes, by a jump discontinuity in the density. Now the plasma hydrodynamic equations which are valid for both plasma media furnish further mathematical boundary conditions. The physical assumptions that are made concern the behaviour of the plasma at the interface and must be consistent with the mathematical conditions (see problem page 209).

We have outlined above possible consistent procedures. Such a procedure has been used successfully to explain quite completely the detailed temperature spectrum of a plasma column (see chapter 5). In some special instances the above approach may prove too coarse. A refined way of stating the boundary conditions would be the following. The distribution function F can be split in F_+ and F_- where F_+ is the distribution function relative to the electrons moving from the plasma towards the interface and F_- is the distribution function relative to the electrons moving away from the interface into the plasma. One can define appropriate moments M_{i+} and M_{i-} involving F_+ and F_- respectively. The physical description at the boundary can be then refined by specifying M_{i+} and M_{i-}. This involves a description by a set of differential equations for the plus quantities and another set of differential equations for the minus quantities. The two sets are not independent, as they are linked by the self-consistent electromagnetic field (**E** and **B**). This approach leads thus to a complicated mathematical description and the added complication will probably only be justified in a few instances. The approach that has just been outlined is, in fact, an extension of the method described by Grad[14].

1.10 Set-ups for studying wave-scattering by plasma columns

Several experimental techniques have been used for studying the

Scattering of Waves by Plasma Columns

scattering behaviour (and in particular the resonances) of plasma columns. These include free-space experiments in which the plasma column is placed in front of a horn or a waveguide acting as an antenna and either the transmitted or the reflected signal is analysed. In another type of experiment the plasma column is inserted perpendicularly to the waveguide and again the transmitted or the reflected signal is analysed. Finally, in another method of observation, the plasma column is placed between a strip-line or specific multipole electrode configurations and the power of the reflected signal is examined. In most cases the frequency of the incident signal is kept constant and the plasma density is varied by changing the discharge current in the column. In some instances, however, the density is kept constant and the frequency is varied. In chapters 3, 4 and 5 extensive references to the experimental work pertaining to each study are given.

a. Method of measurement of a reflection factor in free space

Whatever the set-up employed, most of the experimenters have carried out observations with a quadratic detecting system and special precautions must be taken to compare quantitatively the measured powers corresponding to each resonance. The analysis of the transmitted signal furnishes interesting information, but it must be emphasized that only reflexion results can be quantitatively compared to theory (at least where amplitudes are concerned) because in transmission the observed field is a combination of both the incident and the scattered one. It is thus clear that for a direct comparison between theory and experiments, there is considerable advantage in carrying out reflexion experiments and utilizing a linear detecting system whenever possible.

The amplitude of the scattered field in reflexion as a function of plasma density is the principal study (both experimental and theoretical) in chapters 3, 4 and 5. We will therefore analyse in detail the measurement of a reflexion factor in free space with a *linear* detecting system and follow the analysis of Messiaen and Vandenplas[44]. It is well known that such measurements are difficult to interpret. Conditions must be found under which the interaction between the measuring apparatus and the plasma does not disturb the quantity which must be measured and the precise meaning of this measurement must also be ascertained.

Consider the experimental set-up sketched in figure 1.2. The plasma column of elementary length (length sufficiently small as to allow the waves scattered by different parts of the column to be in phase when entering the antenna) is placed before the antenna (waveguide or horn). The axis of the column is horizontal and perpendicular to the direction of propagation of the wave and to the vertical electric field. We will establish a connexion

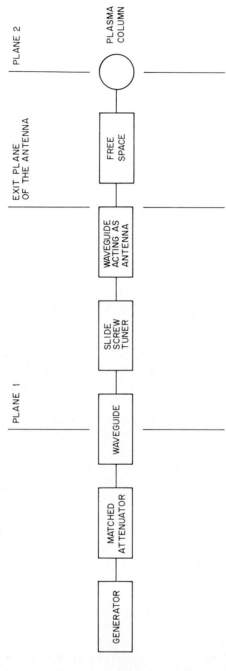

Figure 1.2 Block-diagram of the experimental set-up

between the electromagnetic wave scattered by the plasma column towards the antenna (back-scattering) and the measurements made in the waveguide. Let us consider a plane 1 perpendicular to the symmetry axis of the waveguide and a plane 2, parallel to plane 1, containing the axis of the column. This plasma column is assumed concentrated on its axis (radius of the column is small compared to the vacuum wavelength) and is characterized by an equilibrium electron density $\langle N \rangle$ which is a function of the radius.

The easiest method of calculation consists in considering the multiple reflexions undergone by a signal sent by the generator[15]. The only propagation mode excited in the waveguide by the generator is the TE 10 mode. The waveguide is matched towards the generator by the attenuator. We consider the complex amplitude of the electric field strength of the incident and reflected waves on the symmetry axis of the waveguide; their sum in any given plane is the complex amplitude of the total field. We will examine three distinct cases. The symbol used for all the electric fields travelling from the generator towards the plasma is E_{out}, while the symbol used for all the electric fields travelling from the plasma towards the generator is E_{in}.

Case 1

The column is not present and the matching device (slide-screw tuner) is adjusted in order to match the waveguide towards the antenna. The only wave present in the waveguide is an outgoing wave characterized by the amplitude $|E_{1+}|$ (the index 1 corresponding to plane 1).

Case 2

The position of the slide-screw tuner is the same as in case 1, the only difference being that the plasma column is now present. Let us define the following complex factors:

$$G = \frac{E_{\text{out 2}}}{E_{\text{out 1}}}$$

where $E_{\text{out 1}}$ is the complex value of an outgoing field in plane 1. The definition of $E_{\text{out 2}}$ is similar.

$$T = \frac{E_{\text{out 2}}}{E_{\text{out e}}} = \frac{E_{\text{in e}}}{E_{\text{in 2}}}$$

where the index e stands for the exit plane of the antenna.

$$A = \frac{E_{\text{in 1}}}{E_{\text{in e}}}$$

$$D = \frac{E_{\text{out e}}}{E_{\text{in e}}}$$

This last factor takes into account the scattering effects of both the antenna and the slide-screw tuner.

$$\mathscr{R} = \frac{E_{in\,2}}{E_{out\,2}}$$

$\mathscr{R}$ is thus the plasma column reflexion factor. Note that $\mathscr{R}$ depends on the equilibrium electron density $\langle N \rangle$ while T and G depend on the distance between plasma column and antenna.

Taking for E_{1+} the same definition as in case 1 and applying to it the method of the multiple reflexions (reflexions in planes e and 2, respectively), one obtains the following geometrical series for E_{1-}

$$AT\mathscr{R}GE_{1+}[1 + DT^2\mathscr{R} + \cdots + (DT^2\mathscr{R})^n + \cdots] \qquad (1\text{-}118)$$

This series is easily constructed if one notes that $(E_{in\,e})_{N+1} = T\mathscr{R}TD(E_{in\,e})_N$ where $(E_{in\,e})_N$ and $(E_{in\,e})_{N+1}$ are the incoming fields in the e-plane after N and $N+1$ reflexions on the plasma column respectively. With the attenuation condition $|DT^2\mathscr{R}| < 1$, equation (1-118) gives the total field (in plane 1) which is reflected towards the generator

$$E_{1-} = \frac{AT\mathscr{R}GE_{1+}}{1 - DT^2\mathscr{R}} \qquad (1\text{-}119)$$

It follows that the modulus of the reflexion factor *in the waveguide* has the following form

$$|\Gamma| = \left|\frac{E_{1-}}{E_{1+}}\right| = \frac{|AGT|}{|1 - DT^2\mathscr{R}|}|\mathscr{R}| \qquad (1\text{-}120)$$

This formula shows that $|\Gamma|$ is only proportional to $|\mathscr{R}|$ (which is the quantity we want to measure) when $|DT^2\mathscr{R}| \ll 1$ or, in physical terms, when the interaction between antenna and plasma is sufficiently small. The way to reduce this interaction is to increase the plasma–antenna distance; the latter is also the limit in which incoming TEM waves are obtained in the vicinity of the column.

Case 3

We consider the case when the waveguide is not matched with the antenna in the absence of the plasma column. When there is no column, we define the reflexion factor in plane 1 of the waveguide as being

$$K_1 = \frac{E_{in\,1}}{E_{out\,1}}$$

Scattering of Waves by Plasma Columns

An argument identical to the one followed in case 2 leads to the following formula when the plasma column is present:

$$|\Gamma| = \left| K_1 + \frac{A'G'T\mathscr{R}}{1 - D'T^2\mathscr{R}} \right| \quad (1\text{-}121)$$

where A', D' and G' are not the same as A, D and G because of the alterations introduced by the different position of the screw tuner. The mismatching of waveguide and antenna in the absence of plasma makes it thus impossible to render $|\Gamma|$ proportional to $|\mathscr{R}|$. This is an important conclusion.

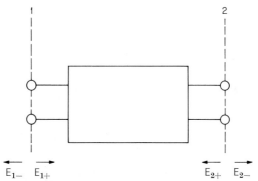

Figure 1.3 Generalized waveguide junction

A second and mathematically more compact method for computing the reflexion coefficient is to consider a generalized waveguide junction with one terminal being the waveguide at plane 1 and the second being the plasma column in plane 2. This generalized waveguide junction thus comprises a section of the waveguide, the screw tuner, the antenna, and free space. The complex amplitudes of incident and reflected fields at the terminals (figure 1.3) are linked together by a scattering matrix **S** such that

$$E_{1-} = S_{11}E_{1+} + S_{12}E_{2+}$$
$$E_{2-} = S_{21}E_{1+} + S_{22}E_{2+} \quad (1\text{-}122)$$

with $S_{21} = S_{12}$ because of the reciprocal character of the network.

Once more we look for the connexion between the reflexion coefficient $\mathscr{R} = E_{2+}/E_{2-}$ of the plasma column and the reflexion coefficient $\Gamma = E_{1-}/E_{1+}$ of the waveguide.

From equation (1-122) we deduce

$$\Gamma = S_{11} + \frac{S_{12}^2 \mathscr{R}}{1 - \mathscr{R}S_{22}} \quad (1\text{-}123)$$

This last formula is identical to (1-121) if one considers the physical meaning of S_{11} (reflexion coefficient in 1 when $E_{2+} = 0$, i.e. in the absence of the plasma column, etc.) The necessity of $S_{11} = 0$ and $|S_{22}| \to 0$, in order to obtain the proportionality of $|\Gamma|$ to $|\mathscr{R}|$ is again obvious.

Experimentally, the direct measurement of $|\Gamma|$ as a function of the equilibrium density $\langle N \rangle$ of the plasma in the column can be made with a

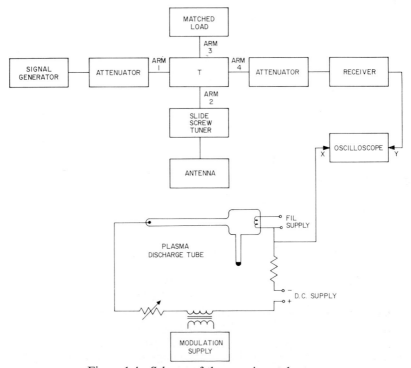

Figure 1.4 Schema of the experimental set-up

magic T used in an impedance bridge (see figure 1.4) or with a directional coupler. The amplitude of the signal obtained in arm 4 is proportional to the modulus of the reflexion ratio Γ in arm 2. This is, of course, only true when the T is not unduly mismatched, and when it is symmetrical. When a UHF–SHF receiver working at a given frequency $\omega/2\pi$, an oscilloscope and a modulated plasma discharge current are used as indicated on figure 1.4, one obtains a curve giving directly $|\Gamma|$ as a function of $\langle N \rangle$ with the assumption that $\langle N \rangle$ is proportional to the discharge current I* in the plasma

* It should be noted that this proportionality constant ($I_{\text{dis}} \simeq -e\langle N \rangle v_{\text{d}}$ where v_{d} is the drift velocity of the electrons along the column) depends slightly on the mean value of the modulated discharge current. Evidently, the ratios of the currents, which often play an important role, are fairly independent of this mean value.

column. A comparison between the direct measurement of $|\Gamma|$ (with a sliding probe) and the amplitude of the Y deviation of the oscilloscope shows that these two quantities are exactly proportional up to $|\Gamma| < 0{\cdot}6$. In most experiments of this type, the plasma is obtained in a mercury discharge tube. Some experiments have also been performed using discharges in other gases (see chapter 5).

b. Detailed resonance spectrum exhibited by a plasma column

As an illustration of the above method, we give some results pertaining to the detailed reflexion spectrum of a cylindrical plasma column (remember that the electric field strength **E** is perpendicular to the axis of the

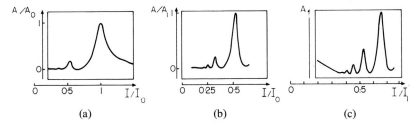

Figure 1.5a Resonance curve giving the amplitude of the reflected E-field as a function of discharge current. (b) Secondary peaks, the leader being the first secondary. (c) Secondary peaks, the leader being the second secondary

column). A typical resonance curve ($\omega/2\pi = 2{\cdot}7$ GHz) is given in figure 1.5a; it displays a main peak to the right and a series of secondary peaks to the left. The amplified secondary peaks are given in figures 1.5b and 1.5c. These latter figures are obtained with considerable amplification of the incident wave and of the reflected wave in the detecting device; the leading peak of figure 1.5b is the first secondary, while the leading peak of figure 1.5c is the second secondary. A_0 and A_1 are the amplitudes of the main resonance and of the first secondary respectively. It will be seen in section 5.2 that the nature and the position of the secondary resonances are perfectly explained by a theory combining non-zero temperature and non-uniform equilibrium density of the plasma. The main peak, as already stated in section 1.4, is quite satisfactorily explained by uniform cold plasma theory.

We now proceed by giving more information about the secondary- or temperature-peaks, but it must be borne in mind that this not only serves

38 *Electron Waves in Bounded Plasmas*

as an illustration of the experimental method but also as an observational introduction to chapter 5.

$$\frac{A_1}{A_0}, \frac{A_2}{A_1}, \frac{A_3}{A_1}, \frac{A_4}{A_1} \quad \text{together with} \quad \frac{I_1}{I_0}, \frac{I_2}{I_1}, \frac{I_3}{I_1}, \frac{I_4}{I_1}$$

are measured as a function of the distance between the antenna and the plasma in free space ($A_0, A_1, \ldots, A_4; I_0, I_1, \ldots, I_4$ are respectively

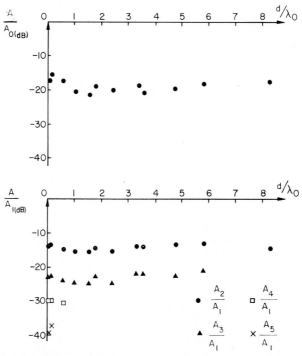

Figure 1.6 Amplitude ratios of the first secondary with respect to the main one and of the successive secondaries with respect to the first secondary as a function of d/λ_0

proportional to the $|\Gamma|$ of the main and secondary peaks and to the currents for which they occur). The results are listed in figure 1.6 for the reflexion ratios and in table 1.1 for the current ratios. They are expressed as a function of d/λ_0 (d is the distance between antenna and plasma column; λ_0 is the vacuum wavelength which is equal to 11·1 cm.) The dispersion of the experimental points on figure 1.6 are due to the sensitive character of the balancing of the impedance bridge by the screw tuner and the noise effects at a great distance.

Scattering of Waves by Plasma Columns 39

The above results were obtained when absorbing screens were put before the discharge tube in such a manner that the reflexions were always due to the same central portion of the plasma. The fact is that the amplitudes observed on the oscilloscope are, notwithstanding the considerable intermediate amplification, rigorously proportional to the intensity of the reflected field, which makes it possible to determine the amplitude ratios even when the distance between the waveguide (acting as antenna) and the plasma column is quite great.

This series of experiments shows that the amplitude and current ratios are roughly independent of d/λ_0 and that they tend toward constant non-zero values when $d/\lambda_0 \to \infty$.

Table 1.1   Current ratios

	Reflexion									Transmission
d/λ_0	0·09	0·18	0·67	1·04	1·57	2·34	3·38	4·85	5·85	8·30	
I_1/I_0	0·520	0·539	0·525	0·530	0·527	0·528	0·525	0·525	0·53	0·54	0·52
I_2/I_1	0·659	0·659	0·652	0·660	0·656	0·652	0·647	0·647	0·66	0·67	0·67
I_3/I_1	0·529	0·530	0·530	0·535	0·527	0·529	0·525	0·52	0·52		0·55
I_4/I_1	0·456	0·456	0·460	0·468	0·459	0·457	0·451				0·47
I_5/I_1	0·405	0·406	0·419	0·418	0·420	0·413	0·413				
I_6/I_1	0·371	0·372									

When transmission experiments are carried out in free space with the same set-up, the current ratios for the different resonances are found, as expected, to be identical to the reflexion ratios. (The receiving waveguide is only 1 cm from the axis of the tube.) The ratios A_1'/A_0', A_2'/A_1', ..., A_4'/A_1' are not exactly complementary to those obtained in the reflexion experiments; this is easily understandable, as the measured field is the combination of the incoming and the scattered fields in contrast with the reflexion case where *only* the scattered field is measured (in the limit where $S_{22} = 0$). The current ratios I_1'/I_0', I_2'/I_1', etc ... (table 1.1) are as expected, identical to the ratios obtained in reflexion.

c. Experiments in the waveguide

When the plasma column is inserted perpendicularly to the waveguide, it distorts the pattern of the fields inside the waveguide and one observes a reflexion spectrum and a transmission spectrum. Several workers have made experiments using this technique and the relevant references will be found in section 5.1. The alterations with respect to the free-space behaviour introduced by the waveguide have been investigated theoretically by Bryant and Franklin using a plasma dielectric description[59,78].

When the other experimental conditions are unchanged, the results corresponding to reflexion experiments with the plasma inside the waveguide show that the current ratios are identical to those obtained in the free-space case but a slightly higher ratio is obtained for A_1/A_0. (Note that since $|\Gamma| > 0.6$ for the main resonance, the main peak should be saturated.)

d. Multipole electrode configurations

In the low frequency region, from approximately 100 MHz to 1·5 GHz the plasma resonances can be observed by placing the plasma column

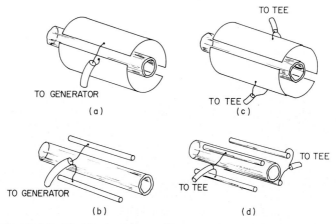

Figure 1.7 Multipole devices. (a) Split cylinder dipole; (b) Wire dipole; (c) Split cylinder quadrupole; (d) Wire quadrupole

between a strip line or in the split-cylinder and wire multipole devices shown on figure 1.7. At resonance, the plasma column absorbs more than the average amount of energy. The radii of the multipole devices are much smaller than λ_0 and the quasi-static approximation is warranted for their analysis[79,81]. If the opposite electrodes in the dipole devices are fed 180° out of phase, the fields produced by them near the axis have a $\cos n\theta$ angular dependence, where $n = 1, 3, 5, \ldots$. However, if the radius of the device is roughly three times larger than the radius of the column, only the dipolar mode ($n = 1$) is appreciably excited. If the adjacent electrodes of the quadrupole devices are fed 180° out of phase, the fields near the axis have a $\cos n\theta$ angular dependence where $n = 2, 4, 6, \ldots$. When the radius of the device is roughly two times larger than the plasma column, the quadrupolar mode ($n = 2$) is preponderantly excited. With multipolar

Measurement of Electron Plasma Densities 41

devices, care must be taken in order to insure that adjacent elements are driven 180° out of phase. For the quadrupole set-up, this condition is obtained by driving opposite elements with two sections of coaxial cables of equal length which are fed from a 'tee'.

1.11 Measurement of electron plasma densities by cavity perturbation techniques

For the experiments described in the later chapters, it is often useful to have an independent measurement of the electron plasma density. Such a measurement can be obtained by using a microwave cavity as proposed by Brown[16,83–89]. The basic idea of this technique is very simple: the plasma column which is being studied is introduced in the cavity and the resulting shift in the resonant frequency and in the Q-factor of the cavity are measured. Usually a first-order perturbation theory is used to interpret these measurements.

For low plasma densities ($\omega_e^2 \ll \omega^2$), for small collision frequencies ($\nu \ll \omega$) and in the absence of a steady magnetic field, perturbation theory leads to the two expressions

$$\frac{\Delta\omega}{\omega_0} \cong \frac{1}{2(1+(\nu/\omega)^2)} \frac{\int_V (\omega_e^2/\omega^2) E_0^2 \, dV}{\int_V E_0^2 \, dV} \qquad (1\text{-}124)$$

$$\Delta\left(\frac{1}{Q}\right) \cong \frac{\nu/\omega}{1+(\nu/\omega)^2} \frac{\int_V (\omega_e^2/\omega^2) E_0^2 \, dV}{\int_V E_0^2 \, dV} \qquad (1\text{-}125)$$

for the frequency shift of the resonance and for the shift in $1/Q$ due to the introduction of the plasma into the cavity. The 0 subscript represents unperturbed values and V is the volume of the cavity.

As an example, we give the theoretical results (see figure 1.8) for a cylindrical cavity excited in the TM_{020} mode as obtained by Brown[86]. The perturbation solution is compared to the exact one assuming a uniform plasma cylinder. The results are also valid when an axial magnetic field $\mathbf{B}_0$ is present, since this field does not affect a mode having only an axial component of the electric field.

The presence of a static magnetic field $\mathbf{B}_0$ usually renders both the theory and the experiments more difficult because the plasma is then anisotropic.

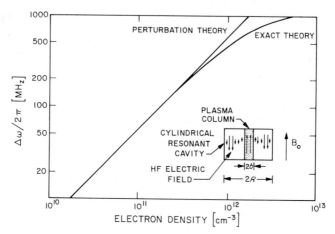

Figure 1.8 Frequency shift of TM_{020} cylindrical cavity with a plasma column; $b/R = \frac{1}{10}$, $\omega_0/2\pi = 5{\cdot}555$ GHz (after reference 86)

Perturbation methods have been investigated in this case and exact solutions have been obtained for some systems with no azimuthal variations[87]. Further work has been carried out[88] and modes not directly related to modes in an empty cavity have also been studied. Corrections due to changes introduced by the holes in the cavity have been calculated by Thomassen[89]. We should also mention recent work by Leprince[221,222] on the perturbation of cavity modes due to the presence of a plasma and on plasma modes of a cavity.

CHAPTER 2

The Uniform Plasma Slab-Condenser System

This chapter deals with the system consisting of a plasma slab of uniform density and a variable region of vacuum between two infinite condenser plates (figure 2.1.). The hydrodynamic theory of this model ($T_e = 0$ and $T_e \neq 0$) is, within certain limits, confirmed by experiments; and it also accounts for the behaviour of the HF plasmoid slab. Finally, the calculation of the impedance of a plasma slab is done using the Boltzmann–Vlasov equation. It is seen that the previous $T_e \neq 0$ results are re-obtained in the appropriate limit excepting, of course, the collisionless damping.

This simple model provides a good introduction to the problems raised by the more complex models, which are studied in later chapters, in the cold and hot plasma limit. It further provides a good heuristic basis for understanding the metallic resonance probe and also, partially, the dielectric resonance probe.

2.1 Theory

The analysis given here, with the exception of section 2.1d, follows rather closely that of Vandenplas and Gould[21,22,23].

a. Equations of the plasma slab-condenser system

The uniform plasma is placed in a container of width a with infinitely thin walls (the effect of finite wall thickness will be investigated later); both other dimensions are infinite. The container is placed between two infinite plane condenser plates P_1 and P_2 and a difference of potential $V(t)$ is applied between them. $E(x)$ is the electric field in the plasma and $E_{\text{ext}}(x)$ is the field outside the plasma but between P_1 and P_2. The distance $P_1 P_2$ is equal to l.

The ions are assumed fixed; the first two linearized hydrodynamic equations for the electrons (see chapter 1) are:

$$\frac{\partial n}{\partial t} + \langle N \rangle \frac{\partial u}{\partial x} = 0 \qquad (2\text{-}1)$$

$$m\langle N \rangle \frac{\partial u}{\partial t} = -e\langle N \rangle E - \gamma KT \frac{\partial n}{\partial x} - \nu m \langle N \rangle u \qquad (2\text{-}2)$$

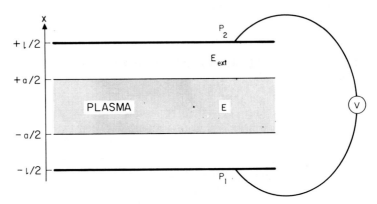

Figure 2.1   The plasma slab-condenser system

γ is a coefficient depending on the assumptions made for the perturbation law of the electronic gas, $\nu m \langle N \rangle u$ is the collision term, and u is the mean macroscopic velocity in the x-direction (in the two other directions this velocity is zero because of symmetry). The translational symmetry along y and z implies $\partial/\partial y = \partial/\partial z \equiv 0$.

The Maxwell equations give:

(i) $$-\langle N \rangle eu + \epsilon_0 \frac{\partial E}{\partial t} = J(t) = \epsilon_0 \frac{\partial E_{\text{ext}}}{\partial t} \qquad (2\text{-}3)$$

because div curl $\mathbf{H}$ = div $[\mathbf{j} + \epsilon_0(\partial \mathbf{E}/\partial t)] \equiv 0$ and the total current density in the x-direction is denoted by J. We notice immediately that E_{ext} is independent of x.

(ii) $$\int_{-l/2}^{-a/2} E_{\text{ext}} \, dx + \int_{-a/2}^{a/2} E \, dx + \int_{a/2}^{l/2} E_{\text{ext}} \, dx - V(t) = 0 \qquad (2\text{-}4)$$

because the quasi-static approximation is made. This approximation is warranted, see section 1.3, because l is much smaller than the vacuum wavelength of the field.

(iii) $$\frac{\partial E}{\partial x} = \frac{-en}{\epsilon_0} \qquad (2\text{-}5)$$

These equations are not independent since (2-3) and (2-5) combine to give (2-1).

It should be noted that the condition $\partial/\partial y = \partial/\partial z \equiv 0$ resulting from our infinite plasma model implies curl $\mathbf{H} \equiv 0$. Therefore, strictly speaking, $\mathbf{J} \equiv 0$ as in any one-dimensional infinite condenser problem. This is not a real difficulty and we assume a slight departure from the ideal model (the

Theory

slab is now very large but not infinite); $\partial/\partial y$ and $\partial/\partial z$ are no longer identically zero and **J** is thus $\neq 0$. This is precisely what is mathematically expressed by equation (2-3).

We must, however, emphasize the fact that we therefore do not consider the waves as being purely longitudinal. Purely longitudinal waves in a plasma are characterized by $\mathbf{H} \equiv 0$. It then follows from Maxwell's equations that **E** is strictly equal to $-\mathrm{grad}\ \phi$. As curl $\mathbf{H} \equiv 0$ implies $\mathbf{J} \equiv 0$, these waves do not give rise to an electric field in an exterior dielectric medium and will thus generally not be observed in this medium. In our model we precisely want to study the effects of the plasma on the **J** measured outside. It will be seen (section 3.1c, equation 2-34), that there exist some discrete values of ω for which **J** will be zero. These correspond to anti-resonances (infinite impedance) of the plasma-slab condenser system and it is only for these particular values of ω that we have purely longitudinal waves. In the plasma-condenser arrangement these longitudinal waves are therefore detected by the existence of an anti-resonance.

b. Cold plasma ($T = 0$)

Neglecting the term in $\partial n/\partial x$, i.e. considering a plasma with $T = 0$, we perform a Laplace transform, characterized by the variable p with respect to time. The Laplace transforms of the functions considered are distinguished by an index p and the initial values for $t = 0$ by a subscript zero.

$$m\langle N\rangle[pu_p - u_0] = -e\langle N\rangle E_p - \nu m\langle N\rangle u_p \qquad (2\text{-}6)$$

$$-\langle N\rangle e u_p + \epsilon_0[pE_p - E_0] = \epsilon_0[pE_{\mathrm{ext}\,p} - E_{\mathrm{ext}\,0}] = J_p \qquad (2\text{-}7)$$

$$\int_0^a E_p\,\mathrm{d}x + E_{\mathrm{ext}\,p}(l - a) - V_p = 0 \qquad (2\text{-}8)$$

u_p is introduced from (2-6) into (2-7) to obtain E_p and $E_{\mathrm{ext}\,p}$ which are then introduced into (2-8). This substitution gives:

$$J_p = -\epsilon_0 \frac{-V_p + \dfrac{E_{\mathrm{ext}\,0}}{p}(l - a) + \dfrac{\langle N\rangle e}{\epsilon_0}\displaystyle\int_0^a \dfrac{u_0}{\omega_e^2 + p^2 + p\nu}\,\mathrm{d}x + (p + \nu)\displaystyle\int_0^a \dfrac{E_0}{\omega_e^2 + p + p\nu}\,\mathrm{d}x}{(p + \nu)\displaystyle\int_0^a \dfrac{\mathrm{d}x}{\omega_e^2 + p^2 + p\nu} + \dfrac{l - a}{p}} \qquad (2\text{-}9)$$

with
$$\omega_e^2 = \frac{\langle N \rangle e^2}{\epsilon_0 m}$$

We examine a few interesting cases when $E_{\text{ext }0} = u_0 = E_0 = 0$, with no collisions ($\nu = 0$).

(a) $V(t) = A\delta(t)$.

$$J_p = \frac{\epsilon_0 A}{l} \frac{p(\omega_e^2 + p^2)}{p^2 + \left(\frac{l-a}{l}\right)\omega_e^2} \quad \text{because} \quad V_p = A.$$

Thus:

$$J(t) = \frac{\epsilon_0 A}{l^2} \omega_e^2 \cos\omega_e \theta t \quad \text{with} \quad \theta = \sqrt{\left(\frac{l-a}{l}\right)} \quad \text{and} \quad 0 < \theta < 1 \tag{2-10}$$

This response introduces the quantity $\omega_e \theta$ which we call the characteristic frequency of the plasma-condenser system. Its importance will appear clearly in the following pages.

If $l = a$,

$$J(t) = \frac{\epsilon_0 A}{l}[\omega_e^2 U(t) + \delta'(t)]$$

where $U(t)$ is the unit function and $\delta'(t)$ the derivative of the delta function. In this case, the condenser-plasma system behaves exactly like a parallel LC circuit and the plasma has an equivalent coefficient of self inductance which is proportional to ω_e^{-2}.

(b) $V(t) = Ae^{-i\omega t}$ for $t > 0$.

$$J_p = \frac{\epsilon_0 A}{l} \frac{p(\omega_e^2 + p^2)}{(p^2 + \theta^2\omega_e^2)(p + i\omega)} \quad \text{because} \quad V_p = \frac{1}{p + i\omega}$$

$$J(t) = \frac{i\epsilon_0 A}{l(\omega_e^2\theta^2 - \omega^2)}[\omega(\omega_e^2 - \omega^2)e^{-i\omega t} + \omega_e^2(1 - \theta^2)$$
$$\times (i\theta\omega_e \sin\omega_e\theta t - \omega\cos\omega_e\theta t)] \tag{2-11}$$

Both the applied frequency and the characteristic frequency appear in the response. If $\omega = \omega_e$, the characteristic frequency $\omega_e\theta$ appears alone. This indicates that $\omega = \omega_e$ is an anti-resonance of the system. The fact that the terms in $\omega_e\theta$ remain present only means that they are the transients created at $t = 0$ when the potential $e^{-i\omega t}$ is switched on; these transients are not damped since there is no resistive element in the circuit.

(c) $V(t) = Ae^{-i\omega_e\theta t}$ for $t > 0$.

Theory

Formula (2-11) gives then 0/0 and the true value is:

$$J(t) = \frac{i\epsilon_0 A}{2\theta l} [i\omega_e(1 - 3\theta^2) \sin \omega_e \theta t + 2\omega_e \theta^2 \cos \omega_e \theta t$$
$$+ it\omega_e^2\theta(1 - \theta^2)e^{-i\omega_e\theta t}] \tag{2-12}$$

There is a resonance phenomenon and $J(t)$ increases linearly with t. This resonance will receive detailed attention later.

If the collisions are taken into account ($\nu \neq 0$), the poles are given by $p^2 + \nu p + \theta^2\omega_e^2 = 0$ or

$$p = -\frac{\nu}{2} \pm i\theta\omega_e \sqrt{\left(1 - \frac{\nu^2}{4\theta^2\omega_e^2}\right)} \simeq -\frac{\nu}{2} \pm i\theta\omega_e \left(1 - \frac{\nu^2}{8\theta^2\omega_e^2}\right)$$

The features of the response are the same except for the appearance of a damping term $e^{-(\nu/2)t}$ and a slight correction to the characteristic frequency. This correction is the same as the one obtained for an infinite plasma[24]. The plasma slab-condenser system gives rise, in this cold plasma approximation, to a simple and clear physical picture if an equivalent electrical circuit is considered. The vacuum layer and the plasma slab constitute, in fact, two condensers in series having respective capacities (per unit surface) C_v and C_p given by $C_v = \epsilon_0/(l - a)$ and $C_p = \epsilon_p/a$ where ϵ_p is the equivalent permittivity of the cold plasma. It is easy to verify that this equivalent circuit exhibits a resonance for $\omega = \omega_e\theta$ and an anti-resonance for $\omega = \omega_e$.

Problem: Prove the above statement.

The fact that the circuit constituted by two condensers in series exhibits a resonance means that the plasma condenser behaves as an inductive medium. Indeed, $\epsilon_p < 0$ when $\omega = \omega_e[(l - a)/l]^{\frac{1}{2}}$ and the capacity C_p is thus negative. This is a very general feature of *cold* plasmas at resonance; we shall see in the following chapters that the resonance frequencies of cold plasma systems are always such that some of the plasma is characterized by $\epsilon_p < 0$.

c. Warm plasma ($T \neq 0$)

We study the harmonic case $V(t) = Ve^{-i\omega t}$. Noting $E = -\partial\phi/\partial x$, equations (2-2), (2-3) and (2-4) lead to the following equation when collisions are neglected:

$$\left(\frac{\partial^2}{\partial x^2} + k^2\right)\frac{\partial^2 \phi}{\partial x^2} = 0 \tag{2-13}$$

with

$$k^2 = \frac{\omega^2 - \omega_e^2}{\gamma KT} \qquad (2\text{-}14)$$

The general solution of (2-13) is:

$$\phi = Ax + B + C \cos kx + D \sin kx \qquad (2\text{-}15)$$

The boundary condition $u = u_w$ for $x = \pm a/2$ is applied to (2-2) and (2-3) after introducing (2-15). The relevant form of u_w will be discussed later.

(i) $u = u_w$ for $x = \pm a/2$ in equation (2-3):

$$u_w \langle N \rangle e + i\omega\epsilon_0(-A \pm Ck \sin ka/2 - Dk \cos ka/2) = i\omega\epsilon_0 E_{\text{ext}}$$

$$(2\text{-}16) \text{ and } (2\text{-}17)$$

The cases $ka = 2N\pi$ or $(2N+1)\pi$ with $N = 0, 1, 2, \ldots$, will be examined later. Equations (2-16) and (2-17) then give

$$C = 0 \qquad (2\text{-}18)$$

$$-A - Dk \cos ka/2 = E_{\text{ext}} - \frac{u_w \langle N \rangle e}{i\omega\epsilon_0} \qquad (2\text{-}19)$$

(ii) $u = u_w$ for $x = \pm a/2$ in equation (2-2):

$$i\omega m \langle N \rangle u_w = e\langle N \rangle [-A - Dk \cos ka/2] + \frac{\gamma\epsilon_0 KT}{e}(-Dk^3 \cos ka/2)$$

$$(2\text{-}20)$$

Equations (2-19) and (2-20) give

$$D = \frac{e\langle N \rangle}{\gamma\epsilon_0 KT k^3 \cos ka/2}\left[eE_{\text{ext}} - i\omega m u_w\left(1 - \frac{\omega_e^2}{\omega^2}\right)\right] \qquad (2\text{-}21)$$

Introducing A, C, D as functions of E_{ext}, i.e. J, in (2-4) and defining Z as V/J, one obtains the 'impedance' Z (in fact the impedance multiplied by the unit of surface of the condenser plates) of the condenser-plasma system:

$$Z = \frac{-\frac{m\omega_e^2}{k^3\gamma KT}\left[-\frac{lk^3\gamma KT}{m\omega_e^2} + 2\tan ka/2 - ka\right]}{\epsilon_0\left\{\left(\frac{m\omega_e^2 u_w}{eV}\right)\left[a + \left(\frac{\omega^2 - \omega_e^2}{k^3\gamma KT}\right)(2\tan ka/2 - ka)\right] - i\omega\right\}} \qquad (2\text{-}22)$$

Equation (2-22) shows that if u_w is not proportional to E at $x = \pm a/2$, i.e. to J and to V, it acts as a source term. The impedance Z is then a function of the two sources V and u_w. The physical reason for this is clear:

Theory

equation (2-2) shows that u is proportional to E or its derivatives everywhere in the plasma and if this is no longer true at the wall, J is not proportional to V as u_w is then effectively another source of the perturbations created in the plasma.

Now if u_w is not a source, we set $u_w = \psi E_{\text{ext}}$ with ψ being some complex constant. Equation (2-22) becomes

$$Z = \frac{-i m \omega_e^2}{\epsilon_0 \omega k^3 \gamma KT} \left\{ \left[\frac{-lk^3 \gamma KT}{m \omega_e^2} + 2\tan \frac{ka}{2} - ka \right] \right.$$
$$\left. - i \frac{m\omega}{e} \psi \left[\frac{ak^3 \gamma KT}{m\omega^2} + \left(1 - \frac{\omega_e^2}{\omega^2}\right)\left(2\tan \frac{ka}{2} - ka\right) \right] \right\} \quad (2\text{-}23)$$

By looking at the second term in the curly brackets of equation (2-23) we see that ψ, if it is not purely imaginary, will introduce a resistance in the otherwise imaginary impedance. In the plasma itself [see equation (2-2)], u is always $\pi/2$ out of phase with E when the collisions are neglected. (The reader will check that this statement is not true in a non-uniform plasma.) We therefore see that if the physical boundary conditions are such that ψ is not purely imaginary, there is damping and Z has a resistive component.

In the model described, the plasma is contained by infinitely thin dielectric walls. In an experimental plasma, a sheath is set up by the negative charge on the dielectric wall and the electrons are electrostatically repelled (see for example reference 6, chapter 3). This suggests that a quite good assumption would be that of specular reflecting wall, i.e. $u_w = 0$ (see section 1.9b). The assumption $u_w = 0$ is of course less specific than that of specular reflexion of each particle in the sheath. Setting $u_w = 0$ corresponds to setting $\psi = 0$ and equation (2-23) becomes

$$Z = \frac{-i m \omega_e^2}{\epsilon_0 \omega k^3 \gamma KT} \left[-\frac{lk^3 \gamma KT}{m \omega_e^2} + 2\tan \frac{ka}{2} - ka \right] \quad (2\text{-}24)$$

The resonance frequencies are given by the zeros of Z. Putting $x = ka$ and $KT/m\omega_e^2 = r_D^2$ where r_D is the Debye radius, $Z = 0$ is given by the solutions of:

$$\frac{x}{2} + \frac{4\gamma l r_D^2}{a^3}\left(\frac{x}{2}\right)^3 = \tan \frac{x}{2} \quad (2\text{-}25)$$

For $\omega < \omega_e$, $k^2 < 0$; we put $k' = ik$, $x' = ix$ and (2-25) becomes:

$$\frac{x'}{2} - \frac{4\gamma l r_D^2}{a^3}\left(\frac{x'}{2}\right)^3 = \tanh \frac{x'}{2} \quad (2\text{-}26)$$

Looking at figure 2.2 one sees that $\tanh x'/2$ is an increasing function of $x'/2$, that the left-hand side of equation (2-26) goes through a maximum and then decreases, and that at the origin $\tanh x'/2$ is below

$$\frac{x'}{2} - \frac{4\gamma l r_D^2}{a^3}\left(\frac{x'}{2}\right)^3 \quad \text{as} \quad \frac{r_D}{a} \ll 1$$

Therefore, there exists a single solution to (2-26) for x' real and positive. A typical value of $(r_D/a)^2$ is $1/1600$; then taking $4\gamma l/a = 16$, one finds that the solution of (2-26) is given by $x'/2 = 9{\cdot}5$. If, instead of the intersection of

$$\frac{x'}{2} - \frac{4\gamma l r_D^2}{a^3}\left(\frac{x'}{2}\right)^3$$

with $\tanh x'/2$, the intersection with the axis is sought, one finds $x'/2 = 10$. This last remark enables us to write the root of (2-26) to an excellent approximation:

$$\frac{x'}{2} - \frac{4\gamma l r_D^2}{a^3}\left(\frac{x'}{2}\right)^3 = 0 \tag{2-27}$$

Thus, using (2-14), the first resonance frequency of the plasma is given by:

$$\omega_0 \simeq \omega_e \sqrt{\left(\frac{l-a}{l}\right)} = \omega_e \theta \tag{2-28}$$

This is the resonance which was found in the cold plasma limit. In order that the approximate formula (2-28) be valid, one must have $r_D^2/a^2 \ll 1$.

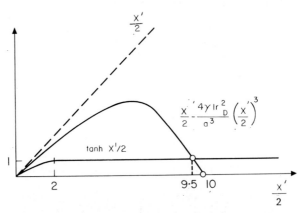

Figure 2.2   Graphical solution of equation (2-26)

Theory

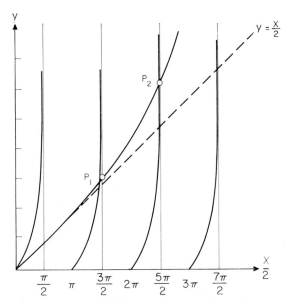

Figure 2.3 Graphical solution of equation (2-25)

When $\omega > \omega_e$, we look graphically for the solutions of (2-25). Figure 2.3 shows that the solutions are approximately given by

$$x_N \cong (2N + 1)\pi \quad \text{or} \quad k_N \cong (2N + 1)\frac{\pi}{a}; \quad N = 1, 2, 3, \ldots \quad (2\text{-}29)$$

The approximation of (2-29) becomes better as N increases. For $N = 1$, $k_1 = (3\pi - 0.19)/a$ instead of $3\pi/a$ for the numerical case considered and the approximation is already quite good. The resonance frequencies are then given by

$$\omega_N^2 \cong \omega_e^2\left[1 + \frac{(2N+1)^2\pi^2\gamma r_D^2}{a^2}\right]; \quad N = 1, 2, 3, \ldots \quad (2\text{-}30)$$

We must now consider the cases when $ka = 2N\pi$ and $ka = (2N+1)\pi$:

(i) $ka = 2N\pi$; $N = 0, 1, 2, \ldots$

Introducing (2-15) in (2-4) leads to

$$Aa - E_{\text{ext}}(l - a) = V \quad (2\text{-}31)$$

Taking into account (2-19), (2-21), (2-31) and $J = -i\omega\epsilon_0 E_{\text{ext}}$, one obtains

$$\frac{a}{i\epsilon_0\omega}\left[\frac{e^2\langle N\rangle}{k^2\epsilon_0\gamma KT} + \frac{l}{a}\right]J = V \quad (2\text{-}32)$$

$Z = 0$ leads then to

$$k^2 = -\frac{a}{l}\frac{e^2\langle N\rangle}{\epsilon_0 \gamma KT}$$

which is impossible for real values of k. Thus, no further resonance frequencies appear for $ka = 2N\pi$.

(ii) $ka = (2N+1)\pi; \quad N = 0, 1, 2, \ldots$

Equations (2-19), (2-20) and (2-4) give:

$$E_{\text{ext}} = -A; \quad A = 0; \quad D = -\frac{V}{2} \tag{2-33}$$

Thus

$$E = -\frac{Vk}{2}\cos kx \tag{2-34}$$

As $E_{\text{ext}} = J = 0$, $Z = \infty$ and we have an *anti-resonance*. We observe standing waves in the plasma with an even number of nodes so that

$$\int_{-a/2}^{+a/2} E\,dx = V$$

This standing wave pattern is a striking feature of the waves in a hot plasma when $\omega > \omega_e$. Furthermore, it should be noticed that the resonances given by the approximate formula (2-30) are narrow as the domain of k between resonance and anti-resonance becomes smaller with increasing N. As noted in the discussion at the end of section 2.1a, these waves are purely longitudinal and are observed as anti-resonances because their existence implies curl $\mathbf{H} = 0$, i.e. $J = 0$.

Let us examine the case when $T \to 0$. Equation (2-30) shows that $(\omega_{N+1} - \omega_N) \to 0$ because $r_D^2 = KT/m\omega_e^2$, that is to say that the resonance frequencies tend toward ω_e. Equation (2-24) shows that the limit ω_e is not a resonance frequency itself. In fact, we have just seen in (ii) that ω_e is an anti-resonance frequency and this remains true when $T_e \to 0$. The isolated resonance frequency will be obtained when $k^2 < 0$. Putting $k = ik'$, observing that $k' \to \infty$ when $T \to 0$ and using (2-24) we see that $\omega \to \omega_e \theta$ because $Z = 0$ gives

$$l = \frac{2\tanh\dfrac{k'a}{2} - k'a}{k'a}\frac{\omega_e^2 a}{\omega^2 - \omega_e^2} \tag{2-35}$$

Problem: Show that for the main resonance of the hot plasma which is given by $\omega = \omega_e \sqrt{\left(\dfrac{l-a}{a}\right)}$ when $r_D \ll a$, the electric field inside the plasma is

$$\frac{E(x)}{E_{\text{ext}}} = \left(1 - \frac{l}{a}\right) + \frac{l}{a}\frac{\cosh\dfrac{x}{r_D}\sqrt{\left(\dfrac{a}{\gamma l}\right)}}{\cosh\dfrac{a}{2r_D}\sqrt{\left(\dfrac{a}{\gamma l}\right)}} \qquad (2\text{-}36)$$

Equation (2-36) clearly shows that $E(x)/E_{\text{ext}}$ goes in a few Debye lengths from 1 at $x = \pm a/2$ to its approximate value of $1 - l/a$ at the centre. This theoretical formula will be compared in section 2.3 to experimental measurements of $E(x)$ in high frequency plasmoids.

d. Effect of collisions

Collisions have been neglected in the previous analysis of a hot plasma. We now investigate the effect of these collisions when they are described in the approximation of a single collision frequency ν as in equation (2-2).

Problem: Show that the impedance Z of the condenser-plasma system is then expressed by

$$Z' = \frac{-\epsilon_0 \omega}{ia} Z = -\frac{l}{a} + \frac{(2\tan ka/2 - ka)\rho\Omega^2}{\gamma k^3 a^3} \qquad (2\text{-}37)$$

where

$$k^2 a^2 = \frac{\rho}{\gamma}[(1-\Omega^2) - i\nu/\omega]; \qquad \Omega^2 = \omega_e^2/\omega^2$$

$$\rho = a^2/r_\omega^2; \qquad r_\omega^2 = KT/m\omega^2$$

$$ka = \left(\frac{\rho}{\gamma}\right)^{\frac{1}{2}}\left\{\left[\frac{[(1-\Omega^2)^2 + (\nu/\omega)^2]^{\frac{1}{2}} + (1-\Omega^2)}{2}\right]^{\frac{1}{2}} - i\sqrt{\left[\frac{[(1-\Omega^2)^2 + (\nu/\omega)^2]^{\frac{1}{2}} - (1-\Omega^2)}{2}\right]}\right\}$$

It will be noted that the parameters entering in equation (2-37) have been chosen in order to study the influence of ν on Z when the frequency $\omega/2\pi$ of the applied V is kept constant and the density of the plasma ($\propto \omega_e^2$) varied.

We compute $|Z'^{-1}|$ as a function of Ω^2 for different values of ν/ω with $\rho = 1000$ and $l/a = 2$. This low value of ρ corresponds to low density plasmas. Such a value of ρ is chosen to render the numerical computations

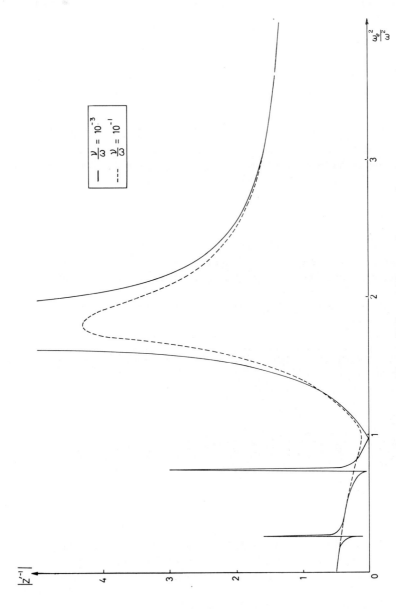

Figure 2.4 $|Z'^{-1}|$ as a function of $\Omega^2 = \omega_e^2/\omega^2$ for $\nu/\omega = 10^{-1}$ and 10^{-3} when $\rho = 1000$ and $l/a = 2$

easier as successive resonances are very close to each other when ρ is high, see equation (2-30). Figure 2.4 shows the results for ν/ω equal to 10^{-1} and 10^{-3}. For $\nu/\omega = 10^{-4}$ the effect of the collisions is found negligible as there exists practically no difference from the results obtained for $\nu = 0$. The main, or 'cold plasma,' resonance occurs at roughly $\Omega^2 = 1\cdot 77$ instead of $\Omega^2 = 2$ as in the $T \to 0$ limit; the influence of the non-zero temperature is rather considerable as ρ is small. The $1/|Z'|$ diagram further exhibits two secondary (or temperature) resonances which are completely damped out for $\nu/\omega = 10^{-1}$. In fact, the dashed curve is quite close to the one obtained for the same value of ν/ω in the $T = 0$ case. The secondary peaks are thus, as expected, very sensitive to ν/ω.

Figure 2.5 displays the details of the effect of the collisions on the temperature peaks. It is clearly seen that in the rather small domain between $\nu/\omega = 10^{-3}$ and $\nu/\omega = 10^{-1}$, the collisions begin by having but little influence on the resonance peaks, and end up by completely washing them out. The temperature peaks of a non-uniform plasma exhibit a comparable behaviour as a function of ν/ω, see reference 65 and chapter 5.

e. Summary

The above results can be summarized as follows:

(1) $T = 0$. The cold plasma-condenser system possesses a single resonance frequency given by $\omega_e\sqrt{((l-a)/l)}$ and a single anti-resonance frequency given by ω_e.

(2) $T \neq 0$. The fundamental characteristic frequency $\omega_e\sqrt{((l-a)/l)}$ is slightly altered if $r_D/a \ll 1$, but a discrete temperature resonance spectrum appears when $\omega > \omega_e$, with frequencies given by

$$\omega_N^2 \cong \omega_e^2 \left[1 + \frac{(2N+1)^2 \gamma \pi^2 r_D^2}{a^2}\right]; \quad N = 1, 2, 3, \ldots \quad (2\text{-}30)$$

The effect of including a non-zero temperature is to remove the degeneracy of a number of eigenfrequencies equal to the electron plasma frequency ω_e. A similar result was obtained by Åström[25] when he studied a plasma slab with electrons of a single velocity in order to investigate the behaviour of a plasma in the region where its equivalent permittivity $\epsilon_p = \epsilon_0(1 - \omega_e^2/\omega^2)$ is equal to zero.

(3) Moderate values of the collision frequency ($\nu/\omega \sim 10^{-2}$ to 10^{-1}) have a relatively small influence on the main resonance, while they strongly damp the temperature resonances.

(4) It will be seen in chapter 5 that the order of magnitude of the spacing between two successive temperature resonances as given by (2-30) is much too small to account for the results of experiments with

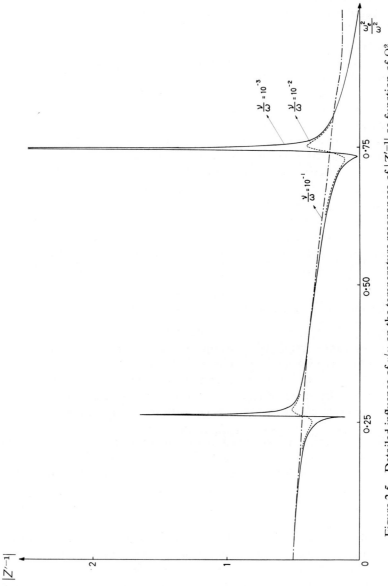

Figure 2.5 Detailed influence of ν/ω on the temperature resonances of $|Z'^{-1}|$ as function of Ω^2

cylindrical plasmas when $(a/r_D)^2 \sim 5 \cdot 10^3$ to 10^4. It will be shown that a satisfactory description of the secondary spectrum must take into account both the non-uniformity and the non-zero temperature of the plasma. However, the present model provides a satisfactory explanation of the experimental situations which will be analysed in sections 2.2 and 2.3.

(5) It will be seen in chapter 6 that this model provides a good heuristic view of the metallic resonance probe.

2.2 Experiments

An experimental set-up was built which was as close an approximation of the theoretical model of section 2.1 as possible (Messiaen and Vandenplas[27]. It was built in such a way (see figure 2.6) that only the current of

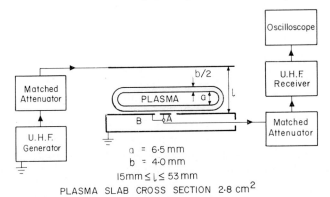

Figure 2.6 Experimental set-up of the plasma-slab condenser system

the central part of the system was measured. This was done in order to approximate the infinite system studied theoretically. We must, however, take the glass walls into account:

Problem: Show that, for a cold plasma system with walls, we have

$$Z = \frac{V}{J} = \frac{-1}{i\omega}\left[\frac{l-(a+b)}{\epsilon_0} + \frac{b}{\epsilon_g} + \frac{a}{\epsilon_p}\right] \quad (2\text{-}38)$$

where $\epsilon_p = \epsilon_0(1 - \omega_e^2/\omega^2)$ is the equivalent permittivity of the cold plasma, and that the characteristic resonance frequency then is

$$\omega_0 = \omega_e \sqrt{\frac{l-(a+b)+\epsilon_0 b/\epsilon_g}{l-b+\epsilon_0 b/\epsilon_g}} \quad (2\text{-}39)$$

with b as defined in figure 2.6, and ϵ_g being the permittivity of the glass.

The plasma is that of a mercury discharge tube with a mercury well at a temperature of 27°C. The oscillographic records are obtained at a given operating frequency and the density of the plasma $\langle N \rangle$ is altered by varying the discharge current ($\langle N \rangle \propto I_{\text{dis}}$). The choice of the frequency is the result of a compromise. The frequency has to be high enough so that the corresponding resonance densities and discharge currents will not be too low, but it cannot become too high, as the corresponding wavelength must remain large with respect to the dimensions of the condenser. The condenser and the plane part of the discharge tube are enclosed in a metallic box to prevent radiation.

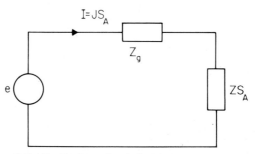

Figure 2.7   Equivalent circuit for the experimental set-up

The equivalent circuit corresponding to the set-up of figure 2.6 is given in figure 2.7. ZS_A is the impedance presented by the central part of the plasma slab-condenser system and Z_g is the impedance resulting from the generator and the other instruments. We have $I = JS_A$ with

$$|I| = \left| \frac{e}{ZS_A + Z_g} \right| \tag{2-40}$$

and we see that Z_g will, in principle, affect the measure of $|Z|^{-1}$. Thus it seems that the anti-resonance ($Z = \infty$), which corresponds to $\omega = \omega_e$, will be more trustworthy than the resonance which corresponds, in the theoretical calculations, to $Z = 0$. The experimentally observed resonance, however, corresponds to a minimum of $|Z_g + ZS_A|$.

The oscillographic records give $|I|$ of (2.40), i.e. practically $|Z|^{-1}$, as a function of the discharge current I_{dis}. In figure 2.8a such a typical record is shown and compared with the theoretical curve of $|Z|^{-1}$ (figure 2.8b) as expressed by equation (2-38). It is to be noted that $|Z|^{-1}$ is plotted versus ω_e^2/ω^2 which is the relevant variable to be compared with I_{dis}.

For a given l, the ratio I_{dis}/ω^2 for resonance and anti-resonance is studied as a function of ω. Theoretically, this ratio should be independent of frequency and it is seen in figure 2.9 that the agreement between theory and experiments is excellent for the anti-resonance and fair for the resonance.

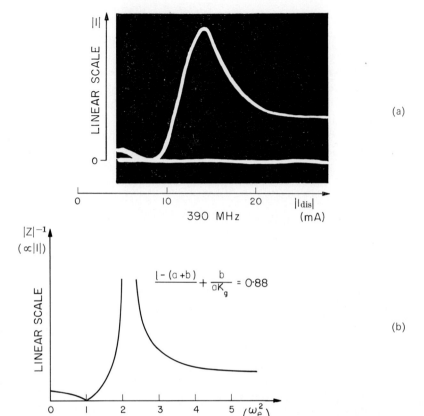

Figure 2.8 The experimental resonance curve giving $|I|$ as a function of $I_{dis} \propto \omega_e^2/\omega^2$ is shown in (a). The theoretical curve giving $|Z|^{-1} \propto I$ as a function of ω_e^2/ω^2 is shown in (b) with $K_g = \epsilon_g/\epsilon_0$

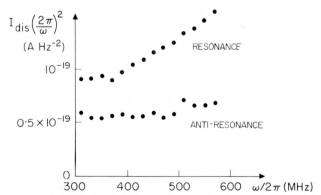

Figure 2.9 Experimental values of I_{dis}/ω^2 versus ω for resonance and anti-resonance

59

The discharge currents corresponding to the resonance and the anti-resonance are now studied for a given frequency as a function of l. The results are summarized in table 2.1. The anti-resonance discharge current is independent of l, as theoretically predicted ($\omega = \omega_e$). It is seen that the experimental behaviour of the resonance as a function of l is only in fair agreement with the theory.

For low frequencies no temperature resonances have been observed. For frequencies higher than 600 MHz, secondary resonances are seen to the left of the main resonance. As the frequency increases, the $|Z^{-1}|$ curve is distorted. This is not surprising, as the wavelength becomes comparable

Table 2.1 $\omega/2\pi = 390$ MHz

$\dfrac{l-(a+b)}{a} + \dfrac{b\epsilon_0}{a\epsilon_g}$	$(I_{\text{dis}})_{\text{anti res.}}$ (mA)	$\dfrac{(I_{\text{dis}})_{\text{res.}}}{(I_{\text{dis}})_{\text{anti res.}}}$	$\dfrac{(\omega_e^2)_{\text{res.}}}{(\omega_e^2)_{\text{anti res.}}}$
0·88	8·1	1·78	2·15
1·30	9·3	1·60	1·76
2·10	8·1	1·52	1·48
3·60	7·9	1·44	1·28
6·70	9·1	1·28	1·15

to the dimensions of the system; the experimental set-up then corresponds less and less to the theoretical model. The secondary peaks are not observed at lower frequencies because the damping effects of the collisions are too strong (see section 2.1d).

The conclusion of this section is clear: the main anti-resonance and the main resonance of a plasma slab-condenser system have been experimentally observed. The agreement with theory, assuming a uniform plasma density, is very good for the former (characterized by $Z = \infty$ for $\omega = \omega_e$ and therefore less subject to the influence of the other impedances of the set-up) and only fair for the latter.

2.3 High frequency plasmoids

High frequency resonance discharges at low pressures, also called high frequency plasmoids, have been known for more than thirty years. They are characterized by a very low exciting HF voltage and an electron plasma frequency of the order of magnitude of the excitation frequency. The physical mechanism responsible for the existence of these plasmoids has been revealed in a beautiful experiment made by Taillet[28,29]. Full bibliographical details concerning these plasmoids are given in the latter reference.

High Frequency Plasmoids

We will briefly describe the experiment, together with the results which are most significant in connexion with the subject matter of this chapter. The basic set-up is sketched in figure 2.10. A vacuum is created in the metallic tank. A discharge having the form of a slab is excited in the residual gas by a high frequency voltage applied to the two plane parallel electrodes. The electric field strength is assumed, in first approximation, perpendicular to the electrodes in the whole of the condenser.

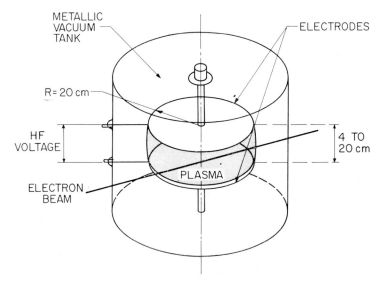

Figure 2.10 Experimental set-up showing the metallic vacuum tank and the two parallel electrodes

The amplitude and the phase of the electric field strength is measured in the following way. The small deviation undergone by an electron beam parallel to the plates, and thus perpendicular to the electric field, is proportional to the amplitude of the field in the region traversed by the beam. If a horizontal sweeping voltage, proportional to the applied HF voltage, is applied to the beam, the resulting trace on a fluorescent screen will give both the amplitude and the phase of the plasma HF electric field strength.

The height a of the plasma slab is measured with a photometer and the electron density with a Langmuir probe. Here one must distinguish the results obtained with electro-negative gases (oxygen, iodine) from those obtained with electro-positive gases (helium, argon, nitrogen). For electro-negative gases, the luminosity of the plasma is practically constant up to the sheaths where the luminosity becomes very low. This is the familiar aspect of plasmoids. The recorded profile shows that the density is practically uniform up to the sheath where it decreases very sharply. One can therefore assume that the sheath region behaves as a vacuum characterized by ϵ_0. This corresponds very well to the theoretical model studied in this chapter. For electro-positive gases, the luminosity, and thus the density, variation from the centre to the edges of the discharge is much more gradual; the theoretical model is more difficult to apply as the 'vacuum' region between plasma and plates is much more arbitrarily defined. The electron beam experiments show that:

(a) The amplitude of the HF electric field, both in the plasma and in the sheath, is roughly ten times greater than that of the field when no plasma is present and when the same voltage is applied between the plates.

(b) The phase of the HF field in the plasma is opposite to the phase of the HF field in the sheath.

This is exactly what is predicted by the theory of section 2.1 for the main resonance at $\omega_e \theta$.

The study of the physical mechanism responsible for the resonant discharge takes into account both the power transmitted by the oscillator to the average electron and the power needed by this average electron in order to sustain the discharge. One is led to the conclusion that there exists a single stable equilibrium value which corresponds to a plasma density characterized by a ω_e slightly greater than ω/θ[29].

When a is measured by the photometer method and ω_e with a Langmuir probe, one can check whether the product $\omega_e \theta$ is equal to the applied frequency. The experimental values found for a and ω_e are such that the frequency of the applied HF voltage is precisely the resonance value $\omega = \omega_e \theta$, the error being of the order of 5 per cent for electro-negative gases. For electro-positive gases, the same result is obtained (ω_e now being, of course, a suitably defined average value) but the error is now of the order of 10 per cent; this last result remains surprisingly good when one thinks of the flexibility involved in the definition of the 'vacuum' region for these electropositive plasmas.

Finally, it is very interesting to compare the spatial distribution of the HF electric field in such a resonance discharge with that predicted by the uniform *hot* plasma theory[30]. The theoretical distribution is given by equation (2-36). Experimental data and theory are compared in figure 2.11. The theoretical curves correspond to different values of a (10, 12 and 14 cm) and thus of ω_e, at the operating conditions: $l = 20$ cm, $\gamma = 3$, $T = 50,000°K$, $\omega/2\pi = 15$ MHz. It must be mentioned that for $a = 10$ cm, the criterion $a \gg r_D$ which must be fulfilled in order that $\omega_{res} \simeq \omega_e \theta$, is but barely satisfied as r_D is of the order of 0·65 cm. As the density does

Impedance of a Plasma Slab Using Vlasov's Equation

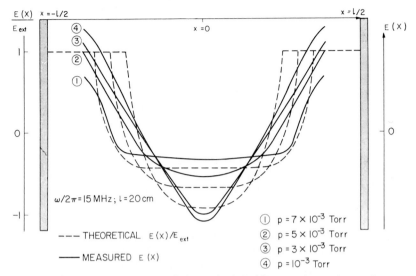

Figure 2.11 Comparison of theoretical $E(x)/E_{\text{ext}}$ and experimental $E(x)$ for different values of the plasma density

not exactly drop to zero in the sheath of the experimental plasma, it is difficult to define an electric field corresponding to the E_{ext} of the model. That is the reason why a non-normalized experimental $E(x)$ is compared to the theoretical $E(x)/E_{\text{ext}}$. Examination of figure 2.11 shows that the phases of the observed electric field, as well as the spatial distribution of this field in the plasma and in the sheath, agree very well with the theoretical curves of the model at $\omega = \omega_e \theta$.

This study of ω_e, a and $E(x)$ demonstrates that the conditions corresponding to the main resonance are attained in the HF discharges or plasmoids. The parameters of the discharge adjust themselves in order that the eigenfrequency of the system be equal to the frequency of the applied voltage: the plasma thus created is resonant.

That this resonantly sustained behaviour is a general feature of plasma systems will be seen for the cylindrical plasma in section 5.4d and for the plasma coated antenna in section 6.3c.

2.4 Impedance of a plasma slab using Vlasov's equation

Several authors have treated the problem of a hot homogeneous plasma slab starting from the linearized Vlasov equation[26,31,32,33,66]. The paper by Shure is very general in scope: it gives a normal mode treatment for the solution of boundary value problems in plasma oscillations and, as an application of the method, the impedance of the plasma-filled parallel plate condenser is calculated when the equilibrium distribution function is Maxwellian. This result is identical to that obtained previously by Gould and we will follow the latter author's treatment closely. Further properties are given in reference 66.

An expression for the impedance per unit area Z of a plasma slab is sought, assuming that (i) The distribution functions of the electrons and the ions both obey Vlasov's equation. (ii) The electrons and the ions are specularly reflected at the boundaries.

The physical situation which we analyse is practically the same as that of section 2.1 when $a = l$, or more correctly, when $E_{\text{ext}} = E_0$. Note, however, that the plasma boundaries are given in the present section by $x = 0$ and $x = a$. The only difference lies in the fact that a uniformly applied electric field E_0 is assumed to drive the plasma and that the motion of the ions is also considered. The impedance Z of the plasma slab is defined as the potential difference between the boundaries divided by the current density, which we found independent of x and equal to $-i\omega\epsilon_0 E_0$. We have

$$Z = \frac{1}{-i\omega\epsilon_0 E_0} \int_0^a E(x, \omega)\,dx \qquad (2\text{-}41)$$

If E has an $e^{-i\omega t}$ time dependence.

The electrons and the ions obey Vlasov's equation

$$\frac{\partial f_e}{\partial t} + v\frac{\partial f_e}{\partial x} - \frac{e}{m_e} E \frac{\partial \langle F_e \rangle}{\partial v} = 0 \qquad (2\text{-}42)$$

$$\frac{\partial f_i}{\partial t} + v\frac{\partial f_i}{\partial x} + \frac{e}{m_i} E \frac{\partial \langle F_i \rangle}{\partial v} = 0 \qquad (2\text{-}43)$$

where $f_e(x, v, t)$ and $f_1(x, v, t)$ are the perturbed one-dimensional electron and ion distribution functions respectively. The electric field is determined by

$$\frac{\partial E}{\partial x} = \frac{e}{\epsilon_0} \int_{-\infty}^{+\infty} (f_i - f_e)\,dv \qquad (2\text{-}44)$$

if the ions have a single positive charge.

a. Boundary conditions

In order to solve equations (2-42) and (2-43) we consider an equivalent problem, periodic in x with period $2a$, as shown in figure 2.12. The electric field just above $x = 2Na$ and just below $x = (2N + 1)a$ ($N = \cdots -2, -1, 0, 1, 2, \ldots$) is E_0. The field is anti-symmetric about each plane $x = Na$; this guarantees that for each particle approaching the plane from above there is a corresponding particle with equal and opposite velocity approaching from below. Thus the condition of specular reflexion is simulated at each of the boundaries. The field E is therefore generated by the surface charges σ_{2N} and σ_{2N+1} indicated in figure 2.12, together with the plasma charge density.

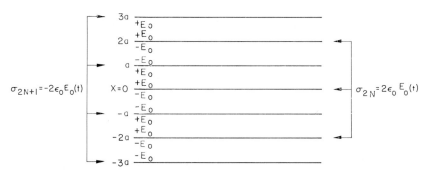

Figure 2.12 The periodic and anti-symmetric arrangement of surface charges and electric fields at $x = 2Na$ and $x = (2N+1)a$

Because of the periodic nature of the problem we Fourier analyse the functions with respect to x:

$$f_e(x, v, t) = \sum_{N=-\infty}^{N=+\infty} f_{eN}(v, t) e^{ik_N x} \quad (2\text{-}45)$$

$$f_i(x, v, t) = \sum_{N=-\infty}^{N=+\infty} f_{iN}(v, t) e^{ik_N x} \quad (2\text{-}46)$$

$$E(x, t) = \sum_{N=-\infty}^{N=+\infty} E_N(t) e^{ik_N x} \quad (2\text{-}47)$$

with $k_N = N\pi/a$.

Furthermore, it should be noted that in solving the periodic problem an extra term appears on the right side of equation (2-44) which represents the surface charges σ_{2N} and σ_{2N+1}:

$$\rho_{\text{source}} = 2\epsilon_0 E(t) \sum_{N'=-\infty}^{N'=+\infty} \{\delta[x - 2N'a] - \delta[x - (2N'+1)a]\} \quad (2\text{-}48)$$

$$\rho_{\text{source}} = \sum_{N=-\infty}^{N=+\infty} \rho_N e^{ik_N x} \quad (2\text{-}49)$$

with

$$\rho_N = \begin{cases} \dfrac{2\epsilon_0 E(t)}{a} & N \text{ odd} \\ 0 & N \text{ even} \end{cases} \quad (2\text{-}50)$$

Let us consider the initial value problem where the field is applied to an undisturbed plasma at $t = 0$. We perform a Laplace transform, characterized by the variable p, with respect to time. This yields

$$(p + ik_N v) f_{eN}(p, v) = \frac{e}{m_e} E_N(p) \frac{\partial \langle F_e \rangle}{\partial v} \quad (2\text{-}51)$$

$$(p + ik_N v) f_{iN}(p, v) = -\frac{e}{m_i} E_N(p) \frac{\partial \langle F_i \rangle}{\partial v} \quad N \text{ odd} \quad (2\text{-}52)$$

$$ik_N E_N(p) = \frac{e}{\epsilon_0} \int (f_{iN} - f_{eN}) \, dv + \frac{2E_0(p)}{a} \quad (2\text{-}53)$$

for each odd Fourier component N. Solving for the Laplace transform of the electric field, we obtain

$$E_N(p) = \frac{2E_0(p)}{iN\pi D(p, k_N)} \quad (2\text{-}54)$$

where

$$D(p, k_N) = 1 + \frac{\omega_e^2}{ik_N \langle N \rangle} \int_{-\infty}^{+\infty} \frac{\partial \langle F_e \rangle / \partial v}{p + ik_N v} \, dv + \frac{\omega_i^2}{ik_N \langle N \rangle} \int_{-\infty}^{+\infty} \frac{\partial \langle F_i \rangle / \partial v}{p + ik_N v} \, dv \quad (2\text{-}55)$$

and ω_e^2 and ω_i^2 are the squares of the electron and ion plasma frequencies respectively. D is the plasma dispersion function of an infinite uniform plasma.

b. Steady state response to $E_0(t) = E_0 e^{i\omega t}$, $t > 0$

If the applied field is given by $E_0(t) = E_0 e^{i\omega t}$ for $t > 0$, then $E_0(p) = E_0 \neq (p + i\omega)$ and we have

$$E_N(p) = \frac{2E_0/iN\pi}{(p + i\omega) D(p, k_N)} \; ; \quad N \text{ odd} \quad (2\text{-}56)$$

The poles of $E_N(p)$ give rise to the contributions to the time-dependent electric field. The contributions from the zeros of $D(p, k_N)$ give rise only however, to exponentially damped terms, so for the steady state response ($t \to \infty$) we need consider only the contribution from the zero of $p + i\omega$. This yields

$$E_N(t) = \frac{2E_0/iN\pi}{D(-i\omega, k_N)} e^{-i\omega t}; \quad t \to \infty \quad (2\text{-}57)$$

Impedance of a Plasma Slab Using Vlasov's Equation

For Maxwellian distributions of electron and ion velocities, the dispersing function $D(p, k_N)$ may be expressed in terms of the derivative of the function

$$Z(z) = \frac{1}{\sqrt{\pi}} \int_{-\infty}^{+\infty} \frac{e^{-x^2} dx}{(x - z)} \; ; \quad \text{Im } z > 0 \tag{2-58}$$

which is tabulated by Fried and Conte[7]. Equation (2-55) becomes

$$D(p, k_N) = 1 - \frac{\omega_e^2}{k_N^2 v_{0e}^2} Z'\left(\frac{-p}{ik_N v_{0e}}\right) - \frac{\omega_i^2}{k_N^2 v_{0i}^2} Z'\left(\frac{-p}{ik_N v_{0e}}\right) \tag{2-59}$$

where $v_{0e} = \sqrt{(2KT_e/m_e)}$ and $v_{0i} = \sqrt{(2KT_i/m_i)}$ are the electron and ion thermal speeds respectively. See chapter 5 of reference 10 for the analysis of Z and D. The function tabulated in reference 7 is defined by (2-58) assuming z to be in the upper half plane; for z in the lower half plane, the analytic continuation is used. Thus the function so defined applies directly to the problem at hand only for *positive* k_N as $p = \pm i\omega + \sigma$ with $\sigma > 0$. For negative k_N, $D(p, k_N)$ is expressed through another function $\tilde{Z}$ which has the same functional form as (2-58) but which is defined in the lower half plane, i.e. for Im $z < 0$. By changing x to $-x$ it can be shown that $\tilde{Z}'(-z) = Z'(z)$ with Im $z > 0$, and this leads immediately to $D(p, |k_N|) = D(p, -|k_N|)$. This last property will be used in establishing equation (2-60).

For the steady state response only a knowledge of Z' for real arguments $(\omega/k_N v_{0e}$ and $\omega/k_N v_{0i})$ is required. From equation (2-47) and using the property stated above, one obtains the total field

$$E(x, t) = \sum_{N=0}^{\infty} \frac{4 \sin \frac{(2N+1)\pi x}{a} E_0 e^{-i\omega t}}{(2N+1)\pi D\left(-i\omega, (2N+1)\frac{\pi}{a}\right)} \; ; \quad t \to \infty \tag{2-60}$$

Integrating the electric field between 0 and a, we have the potential difference

$$-V = \int_0^a E(x, t) \, dx = \sum_{N=0}^{\infty} \frac{8a E_0 e^{-i\omega t}}{(2N+1)^2 \pi^2 D\left(-i\omega, (2N+1)\frac{\pi}{a}\right)} \; ; \quad t \to \infty \tag{2-61}$$

Comparison with equation (2-41) shows that the impedance per unit area is given by

$$Z = \frac{a}{-i\omega\epsilon_0} \sum_{N=0}^{\infty} \frac{8}{(2N+1)^2 \pi^2} \frac{1}{D\left[-i\omega, (2N+1)\frac{\pi}{a}\right]} \tag{2-62}$$

We now examine this expression in detail in some limiting cases. First suppose there is no plasma, $\omega_e^2 = \omega_i^2 = 0$. Then $D = 1$ and since

$$\sum_{N=0}^{\infty} \frac{8}{(2N+1)^2 \pi^2} = 1$$

we have

$$Z = \frac{a}{-i\omega\epsilon_0} = \frac{1}{-i\omega C} \tag{2-63}$$

where $C = \epsilon_0/a$ is the capacitance per unit area.

Next, suppose ω_e^2 and ω_i^2 are not zero, but that the temperature of the plasma tends to zero. The large argument asymptotic expansion of $Z'(z)$ allows us to write in the zero temperature limit

$$D(-i\omega, k_N) = 1 - \frac{\omega_e^2 + \omega_i^2}{\omega^2} \tag{2-64}$$

so that

$$Z = \frac{1}{-i\omega C \left(1 - \dfrac{\omega_e^2 + \omega_i^2}{\omega^2}\right)} \tag{2-65}$$

The impedance is again given by the capacitance formula except that the dielectric constant is now $\epsilon_p/\epsilon_0 = 1 - (\omega_e^2 + \omega_i^2)/\omega^2$. This is the well known cold plasma result already given by equation (2-38), where ω_i^2 is neglected with respect to ω_e^2. There is no dissipation.

Next we discuss the impedance at high frequencies ($\omega \sim \omega_e$) neglecting the ion contributions. Equation (2-62) may be rewritten

$$\frac{-i\omega_e \epsilon_0}{a} Z = \Omega \sum_{N=0}^{\infty} \frac{8}{(2N+1)^2 \pi^2} \frac{1}{1 - \dfrac{\tau^2}{(2N+1)^2 \pi^2} Z'\left(\dfrac{\tau}{\Omega(2N+1)\pi}\right)} \tag{2-66}$$

where $\Omega = \omega_e/\omega$ is a normalized frequency and $\tau = \omega_e a/v_0$ is the thickness of the slab measured in $\sqrt{2}$ times the Debye length. Written this way we see there are two important parameters, the frequency and the slab thickness. We see that if $\tau \gg \pi$ we may employ the large argument approximation to Z' for the first few terms in the series, and that the contribution from large N terms is not significant because of the factor $(2N+1)^2$ in the denominator of equation (2-66). The impedance is expressible as an infinite series. The denominator of each term contains $D(-i\omega, k_{2N+1})$. Thus if the frequency of the driving field coincides with the frequency of a plasma wave whose wave number is $k_{2N+1} = (2N+1)\pi/a$ the denominator vanishes and there is a pole in the impedance function. Using the large argument asymptotic expansion of the function Z', these anti-resonance frequencies are

$$\omega_{2N+1}^2 = \omega_e^2 \left[1 + \frac{3(2N+1)^2 \pi^2 r_D^2}{a^2} \right]; \quad N = 0, 1, 2, \ldots \quad (2\text{-}67)$$

We obtain a result which is identical to that obtained from the moments approach (section 2.1c) if the γ of that section is taken equal to 3. This $\gamma = 3$ value is precisely the one found by a consistent use of the moment approach (section 1.6). This wave-mixing is linked to the reflexion of the particles on the walls of the condenser, the damping term decreasing when a increases.

Equations (2-62) or (2-66), however, also predict a resistive component of the impedance function because the asymptotic expansion of Z' also possesses a small imaginary component. Since the absence of collisions was postulated, some care is needed in its interpretation. One is tempted to attribute the result to a steady state phenomenon (wave-mixing) corresponding in some way to the Landau damping present in the initial value problem[53]. We are dealing with the steady state response where both ω and k are real, so it is impossible to avoid the singularity on the path of velocity integration in equation (2-55). Furthermore, the linearized approximation breaks down for particles which travel at the wave speed ω/k_N. We may, however, consider this problem in another way by taking into account the fact that the electrons do always experience a small collisional damping and by imagining that the applied field is very weak so that individual electrons do not travel at the wave speed for a sufficiently long time for non-linear effects to become important. The assumption of a small collisional damping changes the denominator of equation (2-55) to $p + \nu + ik_N v$ (where ν is the collision frequency) and the previous results described by equation (2-66) now hold for sufficiently small ν and sufficiently weak applied fields. The resistance in the impedance expression (2-66) may then be thought of as due to absorption of energy by particles traveling near the wave speed and its subsequent transfer to other electrons as heat.

This treatment of the hot uniform plasma slab starting directly from the linearized Vlasov equation shows that we obtain, in the cold plasma limit, formula (2-65) which is identical to that derived in the same limit from the fluid description. When $T \neq 0$ and $r_D \ll a$ (Debye length small with respect to the slab thickness), we obtain formula (2-67) which, again, is identical to that derived in the same approximation from the hydrodynamic approach. We therefore see that, with the exception of the damping phenomena, the use of the Vlasov equation leads as expected to the same results as the use of the hydrodynamic equations.

Problem: Show that the impedance of the plasma slab-condenser system of section 2.1 is obtained by merely adding $(l - a)/(-i\omega\epsilon_0)$ to the Z of the present section as defined by equation (2-41).

CHAPTER 3

The Hollow Cylindrical Plasma

We now consider the hollow cylindrical plasma both experimentally and theoretically in the cold plasma approximation. It is shown that two well-identified resonance peaks are observed and that the agreement between experimental and theoretical curves is very good. The classical results of the cold plasma circular column are retrieved by letting the inner radius tend to zero.

A quantitative comparison of the data obtained with the ordinary cylindrical plasma column and with the hollow plasma column shows that the resonances give a very satisfactory measurement of the average plasma density.

When the two cylinders limiting the plasma are not centred on the same axis, it is shown theoretically that each resonance of the hollow plasma splits into a doublet of resonances having a $\cos^2 \theta$ and a $\sin^2 \theta$ angular dependence, respectively. These predictions are in very good agreement with experimental data; thus the cold plasma theory also explains the asymmetry effects very well.

In this chapter we will follow the lines of the work by Messiaen and Vandenplas[34,35,36].

3.1 Theory

An incoming plane wave characterized by its electric field strength $\mathbf{E}_i$ and its magnetic induction $\mathbf{B}_i$ is scattered by a hollow cylindrical plasma column as sketched in figure 3.1.

When the electron temperature and the motion of the ions are neglected and the equations describing the plasma linearized, the plasma can be characterized by an equivalent permittivity (see section 1.2). Furthermore, if the collisions are neglected, this permittivity is real and we shall make the additional assumption that the plasma density is constant. The permittivity ϵ_p of the plasma is given by

$$\epsilon_p = \epsilon_0 \left(1 - \frac{\omega_e^2}{\omega^2}\right) \tag{3-1}$$

with the usual notations. We will use the quasi-static approximation, i.e. take curl $\mathbf{E} = 0$, and this is warranted (see section 1.3) when

$$\left| \epsilon_0 \left(1 - \frac{\omega_e^2}{\omega^2}\right) \mu_0 \omega^2 L^2 \right| \ll 1 \tag{3-2}$$

Theory

where μ_0 is the permeability of vacuum and L is a characteristic length of the problem (here $L = b$; see figure 3.1). The problem is then formally equivalent to an electrostatic problem with $\mathbf{E} = -\nabla\phi$ and we have

$$\text{div } \epsilon_p \mathbf{E} = 0 \tag{3-3}$$

Thus $\nabla^2\phi = 0$ when $\epsilon_p \neq 0$. Particular solutions with separate variables of the Laplace equation in cylindrical coordinates are

$$\phi = (K_{1n}r^n + K_{2n}r^{-n})e^{in\theta}; \quad n \neq 0 \tag{3-4}$$

We consider an exciting field of the form $r^n \cos n\theta$ in the whole of space; because of the quasi-static approximation this represents the incoming

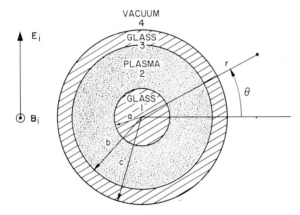

Figure 3.1 Cross section of the hollow cylindrical plasma column with the vectors $\mathbf{E}_i$ and $\mathbf{B}_i$ of the plane incoming wave

field very satisfactorily in the neighbourhood of the column. It should be kept in mind, however, that as $c \ll \lambda_0 (\lambda_0$, vacuum wavelength of the field) the excitation is preponderantly dipolar ($n = 1$), i.e. the electric field seen by the hollow column is practically uniform and is taken in our calculation as being of unit amplitude. The different regions considered are as indicated in figure 3.1; the permittivity of glass is denoted by ϵ_g. In region 1, ϕ_1 must be bounded on the axis, and in region 4 the part of ϕ_4 corresponding to the scattered field has to vanish at infinity. Thus:

$$\phi_1 = A_n r^n \cos n\theta \tag{3-5}$$

$$\phi_2 = B_n r^n \cos n\theta + C_n r^{-n} \cos n\theta \tag{3-6}$$

$$\phi_3 = D_n r^n \cos n\theta + F_n r^{-n} \cos n\theta \tag{3-7}$$

$$\phi_4 = G_n r^{-n} \cos n\theta + r^n \cos n\theta \tag{3-8}$$

The boundary conditions at each surface of separation are given by the continuity of the tangential components of the electric field strength and by the continuity of the normal components of the electric displacement. Since we need these conditions later, we will write them here.

$$A_n a^n = B_n a^n + C_n a^{-n} \tag{3-9}$$

$$B_n b^n + C_n b^{-n} = D_n b^n + F_n b^{-n} \tag{3-10}$$

$$D_n c^n + F_n c^{-n} = G_n c^{-n} + c^n \tag{3-11}$$

$$\epsilon_g A_n a^{n-1} = \epsilon_p (B_n a^{n-1} - C_n a^{-n-1}) \tag{3-12}$$

$$\epsilon_p (B_n b^{n-1} - C_n b^{-n-1}) = \epsilon_g (D_n b^{n-1} - F_n b^{-n-1}) \tag{3-13}$$

$$\epsilon_g (D_n c^{n-1} - F_n c^{-n-1}) = \epsilon_0 (c^{n-1} - G_n c^{-n-1}) \tag{3-14}$$

The amplitude of the scattered field in vacuum, noted G_n, is obtained by solving the system of equations (3-9) to (3-14). Thus

$$G_n = c^{2n} \frac{R_n \epsilon_p{}^2 - \epsilon_p \epsilon_g (R_n + S_n)\left(\dfrac{a^{2n} + b^{2n}}{a^{2n} - b^{2n}}\right) + \epsilon_g{}^2 S_n}{P_n \epsilon_p{}^2 - \epsilon_p \epsilon_g (P_n + Q_n)\left(\dfrac{a^{2n} + b^{2n}}{a^{2n} - b^{2n}}\right) + \epsilon_g{}^2 Q_n}, \quad n \neq 0 \tag{3-15}$$

with

$$\begin{matrix} P_n \\ Q_n \end{matrix} = \epsilon_0 (b^{-2n} \mp c^{-2n}) + \epsilon_g (b^{-2n} \pm c^{-2n})$$

$$\begin{matrix} R_n \\ S_n \end{matrix} = \epsilon_0 (b^{-2n} \mp c^{-2n}) - \epsilon_g (b^{-2n} \pm c^{-2n})$$

The two resonance frequencies for each n are given by the roots ϵ_{p1} and ϵ_{p2} of the denominator of (3-15). The fact that a hollow plasma cylinder possesses two resonances was first observed by Kaiser and Closs[37]. The plasma tube is transparent when $G_n = 0$, i.e. for the roots ϵ_{p1}' and ϵ_{p2}' of the numerator of (3-15).

Let us examine the limit of equation (3-15) when a tends to zero. The amplitude of the scattered field is then given by

$$G_n = c^{2n} \frac{R_n \epsilon_p + S_n \epsilon_g}{P_n \epsilon_p + Q_n \epsilon_g} \tag{3-16}$$

and, in this case, one obtains a single resonance frequency. Equation (3-16) is identical to the one obtained when treating the ordinary plasma cylinder; this is easily seen, using equations (3-9) to (3-14), by setting $A = C = a = 0$. The scattering behaviour of the ordinary plasma column is thus the limiting case of the scattering behaviour of the hollow plasma column when the inner radius a tends to zero.

Problem: Show that the same formula (3-15) for the amplitude G_n of the scattered field is obtained if one uses the full electromagnetic approach, that is, the approximation $\mathbf{E} = -\nabla \phi$ is not made, and if the asymptotic formulas for $kr \ll 1$ of the Bessel, Neumann and Hankel functions

$$J_n(kr) \simeq \frac{1}{n!} \left(\frac{kr}{2}\right)^n \qquad (3\text{-}17)$$

$$N_n(kr) \simeq -\frac{(n-1)!}{\pi} \left(\frac{2}{kr}\right)^n; \qquad n \neq 0 \qquad (3\text{-}18)$$

$$H_n^{(1)}(kr) \simeq iN_n(kr) \simeq -\frac{i(n-1)!}{\pi} \left(\frac{2}{kr}\right)^n; \qquad n \neq 0 \qquad (3\text{-}19)$$

are used. Approximations (3-17), (3-18) and (3-19) are equivalent to the quasi-static approximation. Approximation (3-19), where the real part of the Hankel function of the first kind is neglected with respect to its imaginary part, renders the fields infinite at resonance, as it is this real part which is responsible for the radiation damping of the scattered field.

Problem: Show that the $n = 0$ contribution to the scattered field is zero if $|kr|^2 \ll 1$ where the boundary conditions are stated.

Problem: Show that for a cold uniform hollow plasma of equivalent permittivity ϵ_p placed in vacuum with no glass wall ($b = c$; $\epsilon_g = \epsilon_0$), the two dipolar resonances ($n = 1$) are given by

$$\epsilon_{p1} = -\epsilon_0 \frac{b+a}{b-a} = -K\epsilon_0; \qquad \epsilon_{p2} = -\epsilon_0 \frac{b-a}{b+a} = -\epsilon_0/K \qquad (3\text{-}20)$$

Show furthermore, that the amplitudes of the corresponding surface charge densities σ_a and σ_b at $r = a$ and $r = b$ (see section 1.9a), which become infinite at resonance, are characterized by the ratios

$$\left(\frac{\sigma_b}{\sigma_a}\right)_{p1} = \frac{a}{b} ; \qquad \left(\frac{\sigma_b}{\sigma_a}\right)_{p2} = -\frac{a}{b} \qquad (3\text{-}20 \text{ bis})$$

The latter results give a good pictorial representation of the two eigenstates of the hollow plasma which correspond to the two eigenfrequencies of the problem. It is suggested that the interested reader draw the electric lines of force in both cases.

3.2 Comparison between experiments and theory

The experimental set-up is the one described in section 1.10a for reflexion experiments with a plasma column placed outside a waveguide. We will stress only that, because of the linear detecting system used, the modulus of the reflexion factor is displayed on the oscilloscope screen as a function of the discharge current in the plasma tube. This reflexion factor is the ratio of the back-scattered field to the incoming plane wave field and, as $c \ll \lambda_0$, it corresponds to the G_1 of equation (3-15) with the dipolar ($n = 1$) exciting field of unit amplitude.

The inner glass rod of figure 3.1 is centred with a precision of $\pm 0\cdot 1$ mm for a length of 20 cm. The plasma is obtained in the positive column of a mercury discharge tube with a mercury well at a temperature of 27°C. The permittivity ϵ_g of the glass used is $4\cdot 3\ \epsilon_0$ at the operating frequency of 2·7 GHz.

The experiments were carried out with two different hollow tubes. The geometrical characteristics of the tubes are given in table 3.1. Since in

Table 3.1

TUBE	a (mm)	b (mm)	c (mm)	Theoretical $(\omega_{e1}/\omega_{e2})^2$	Experimental $(\omega_{e1}/\omega_{e2})^2 = J_1/J_2$
1	1·10	3·97	5·25	0·36	0·42
2	1·45	3·80	5·27	0·27	0·32

the experiment ω is kept constant and the equilibrium electron density $\langle N \rangle$, i.e. $\omega_e^2 = \langle N \rangle e^2/m\epsilon_0$, is varied, the theoretical curves for $|G_1|$ were plotted as a function of ω_e^2/ω^2. This permits immediate comparison with the experimental curves where $|G|$ is given as a function of the average current density J because

$$J = -|e|\langle N \rangle v_d = -\frac{m\epsilon_0}{|e|}\omega_e^2 v_d$$

The drift velocity of the electrons is v_d and the motion of the ions is, of course, neglected.

The experiments must be performed in a very short time so as to prevent excessive heating of the inner rod. This rod is heated by the recombination energy of ions and electrons as well as by collisions; because of the very small thermal conductivity of the glass and the lack of external cooling this heat cannot be evacuated. After a time interval of the order of 30 sec, the distortion of the resonance curve shows that the thermal deformations of the central rod (which is fixed at both ends) are then important enough to alter the geometrical characteristics of the system.

The theoretical and experimental curves for tube 1 are given in figures 3.2a and 3.2b respectively. Those of tube 2 are given in figures 3.3a and 3.3b. In table 3.1 are also listed the theoretical values of $\omega_{e1}^2/\omega_{e2}^2$ corresponding

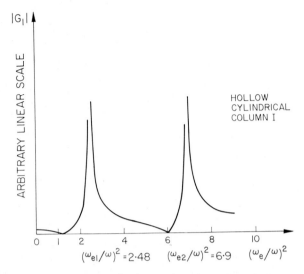

Figure 3.2a Theoretical $|G_1|$ as a function of $(\omega_e/\omega)^2$ for tube 1

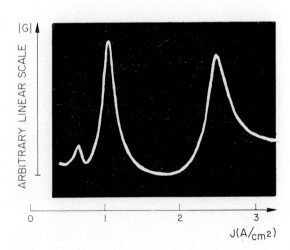

Figure 3.2b Experimental $|G|$ as a function of $J(\propto \omega_e^2)$ for tube 1

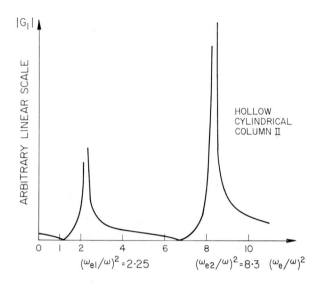

Figure 3.3a Theoretical $|G_1|$ as a function of $(\omega_e/\omega)^2$ for tube 2

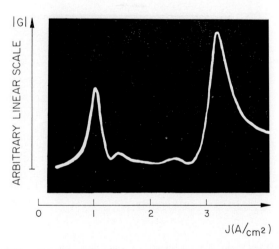

Figure 3.3b Experimental $|G|$ as a function of $J(\propto \omega_e^2)$ for tube 2

to the two resonance peaks as well as the experimental values of $\omega_{e1}^2/\omega_{e2}^2 = J_1/J_2$*.

At the resonance frequencies the theoretical $|G_1|$ is infinite because collisions and radiation damping have been neglected. The curves have been drawn limiting arbitrarily $|G_1|$ at a given value of $\Delta(\omega_e/\omega)^2 = 0.1$ around the resonance values so as to get a good idea of the relative amplitudes†. It should be noted that the relative theoretical amplitudes agree with those obtained experimentally and this agreement is found for both discharge tubes.

Secondary resonances are also observed on the oscillographic records and are seen, for example, very clearly on the left of the main peak of figure 3.4b. These secondary resonances are due to the non-zero electron temperature. They have already been investigated in chapter 2 for a homogeneous plasma and will be thoroughly treated in chapter 5 when the plasma is inhomogeneous. The small bump situated on the right of the main resonance (figure 3.4b) is a $n = 2$ cold plasma resonance, as will be clearly shown in section 4.3. Its position agrees very well with theory.

It must be emphasized that the close agreement between theory and experiments as shown in table 3.1 and figures 3.2a, 3.2b and 3.3a, 3.3b is rather rare in plasma physics. In fact, one could be surprised that a theory assuming constant equilibrium electron density fits so closely the experimental results when it is well known that this plasma density is not radially constant in the discharge tube. To understand this one should remember the results of our general discussion in chapter 1 of the equivalent permittivity approach. It was shown that very important and rather precise results are obtained by using a constant plasma density equal to the average radial plasma density but that the use of a variable density had intrinsic limitations in the $T = 0$ approach. For example, if $\langle N \rangle = \langle N_0 \rangle [1 - \alpha(r/b)^2]$ with $\alpha \leq 0.6$ in a cylindrical plasma, this constant density approach, using the average density in the column, yields the same resonance frequency (with an approximation of 2.5 per cent) as that obtained by solving $\nabla(\epsilon_p \nabla \phi) = 0$ with ϵ_p having the parabolic radial dependence. It should be remembered, furthermore, that the resonance frequency of a uniform cold plasma is practically identical to the main resonance frequency of a hot uniform plasma only if $r_D \ll b$ (see section 2.1d).

With these restrictions of $r_D/b \ll 1$ and of not departing too strongly from constant density in the bulk of the plasma, two slightly different

---

* This assumes that v_d is independent of J for a given tube. This assumption is experimentally verified for cylindrical discharge tubes. It is seen that when working on a given column with different frequencies ω, the resonance current densities J are proportional to ω^2 in a rather wide domain and this is, of course, only possible if v_d remains practically independent of J.

† That this is a satisfactory procedure is seen from sections 2.1d and 5.2e.

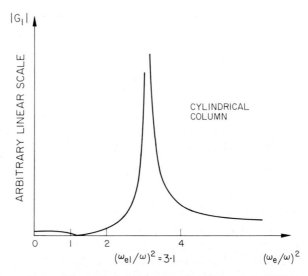

Figure 3.4a Theoretical $|G_1|$ as a function of $(\omega_e/\omega)^2$ for the cylindrical tube

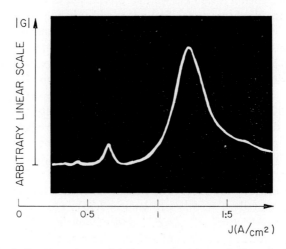

Figure 3.4b Experimental $|G|$ as a function of $J(\propto \omega_e^2)$ for the cylindrical tube

resonant frequencies can be obtained by taking a radial density distribution into account. It is not at all certain, however, that these frequencies are more precise than those obtained by using a uniform density equal to the average density in the column (see sections 1.4 and 1.5).

Problem: Show by the variational method that if the potential (3-6) of the uniform hollow plasma is used as a trial function and if the ensuing expression for ω_0^2 is minimized with respect to the parameter B_n/C_n, the two following resonant frequencies are obtained for the n-th mode of the non-uniform hollow plasma (see reference 38).

$$\omega_{n\pm}^2 = \frac{1}{\left(\frac{J_1}{I_1} + \frac{J_2}{I_2}\right) \pm \left[\left(\frac{J_1}{I_1} - \frac{J_2}{I_2}\right)^2 + \left(\frac{a}{b}\right)^{2n} \frac{J_3^2}{I_1 I_2}\right]^{\frac{1}{2}}} \qquad (3\text{-}20)$$

where

$$I_1 = \frac{\int_a^b \omega_e^2(r) \frac{d}{dr}\left(\frac{r}{b}\right)^{2n} dr}{1 - (a/b)^{2n}} \quad ; \quad I_2 = \frac{\int_b^a \omega_e^2(r) \frac{d}{dr}\left(\frac{a}{r}\right)^{2n} dr}{1 - (a/b)^{2n}}$$

$$J_1 = \frac{1}{2}\left[1 + \frac{\epsilon_g/\epsilon_0}{1 - (a/b)^{2n}}\left(\frac{Q_n}{P_n} + \left(\frac{a}{b}\right)^{2n}\right)\right]$$

$$J_2 = \frac{1}{2}\left[1 + \frac{\epsilon_g/\epsilon_0}{1 - (a/b)^{2n}}\left(1 + \frac{Q_n}{P_n}\left(\frac{a}{b}\right)^{2n}\right)\right]; \quad J_3 = \frac{\epsilon_g/\epsilon_0}{1 - a/b}\left(1 + \frac{Q_n}{P_n}\right)$$

3.3 Measurement of average plasma densities by the main resonances of cylindrical structures

It is interesting to try to compare quantitatively the theoretical and experimental results (see figure 3.4a and 3.4b) obtained with a cylindrical plasma column ($b = 3\cdot97$ mm; $c = 5\cdot25$ mm) to those obtained with the two hollow tubes. This comparison can be done easily as the results are given as a function of the current density J and not the total current I. The different values of the experimental J and the theoretical ω_e^2/ω^2 corresponding to resonance are listed in table 3.2.

As previously noted, there is experimental evidence that, for a given discharge tube and a given vapour pressure, the drift velocity v_d is roughly constant. This enables us to make the comparison between theory and experiment listed in table 3.1. In table 3.2 we normalize the experimental J by considering that there is perfect agreement between theory and experiment in the case of the cylindrical column. The normalized J for the cylinder thus has the $3\cdot10$ value of the theoretical $(\omega_e/\omega)^2$. This procedure is in fact a determination of the v_d of this cylindrical column, as we assume that the experimental $\langle N \rangle$ and the theoretical ω_e^2 correspond perfectly.

The normalization factor determined in this fashion also normalizes the resonance current densities of the hollow tubes. The fact that these normalized values of J correspond quite closely to the theoretical values of $(\omega_e/\omega)^2$ for the hollow cylindrical tubes 1 and 2 shows that the values of v_d for these tubes are about the same as the value of v_d for the cylindrical column. Note that this is not unexpected because the mercury vapour pressure is the same in the three cases and the area of the cross-sections is comparable; under such conditions the DC axial electric field in the three

Table 3.2

	Experimental J (A/cm^2)		Normalized J $J_{\text{cyl}} = 3 \cdot 1$		Theoretical $(\omega_e/\omega)^2$	
Resonance	1	2	1	2	1	2
Cylindrical tube	1·23		3·10		3·10	
Tube I	1·04	2·47	2·60	6·20	2·50	6·90
Tube II	1·02	3·20	2·55	8·00	2·25	8·30

columns are approximately equal and equation (2-26) of reference 8 shows that the corresponding three drift velocities are theoretically indentical.

In addition, the close agreement obtained with different cylindrical and hollow cylindrical tubes, between the theoretical and experimental values of the resonant frequencies, demonstrates that the measurement of the main resonant frequency (or frequencies) is a rather precise method for determining the average plasma density in such columns. Later work by Crawford and others[39] shows that the density obtained by the above method is in quite fair agreement with the density obtained by measurements with the Langmuir probe and the cavity perturbation technique. Further, it must be remarked that the use of the main resonance of a cylindrical plasma as a method of measurements of average plasma density was first suggested by Boyd[42].

3.4 Asymmetry effects in a hollow cylindrical plasma

When the central glass rod is not perfectly centred with respect to the outer glass cylinder, it is observed that instead of the two aforesaid resonance peaks (sections 3.1 and 3.2) of the hollow plasma column, two pairs of resonance peaks appear; each pair growing out of one of the two resonance peaks characterizing the centred case. This phenomenon must be pictured in the following way. The asymmetry gives rise to a preferred direction in the cross-section of the plasma column along the line joining the centres of both cylinders. Because of the existence of this preferred direction, the $\cos \theta$ and $\sin \theta$ modes are no longer degenerate: each of the former peaks gives rise to a resonance doublet. Each component of a resonance doublet will have its own angular dependence.

We shall investigate these asymmetry effects theoretically. We make the same theoretical assumptions as in section 3.1 and we know that the problem is then formally equivalent to an electrostatic problem with $\mathbf{E} = -\nabla\phi$ and $\nabla(\epsilon_p \nabla\phi) = 0$. The natural coordinate system for problems involving two non-concentric cylinders is the biaxial (or bipolar) one and it must be noted that the Laplace equation separates in this system.

A first calculation is performed with a non-centred hollow cylinder of plasma in vacuum and is extended in a second calculation to take the glass walls into account. The fields in the different media are described by potentials which are solutions of the Laplace equation in the biaxial coordinate system. This involves a mathematical difficulty, as the particular solutions of the Laplace equation which have the correct behaviour at infinity display a singularity in a point at finite distance which is inherent in the coordinate system used. It is shown, however, that the form for the scattered field is, for small eccentricity, asymptotically correct in the region near the plasma tube, i.e. where the boundary conditions between the different fields are stated. It is noted that the solution obtained is in fact rigorously correct if the behaviour of two identical asymmetrical hollow plasmas placed apart at a well-defined distance is studied.

Very good numerical agreement is found between the four theoretical resonant frequencies and those observed experimentally[35]. The $\cos \theta$-dependence (θ is the polar angle in the cross-section of the cylinders as shown in figure 3.6) of one component and the $\sin \theta$-dependence of the other component of a resonance doublet is completely confirmed by experiments. It is thus seen that the asymmetry effects of a hollow cylindrical plasma are well understood by a cold plasma analysis.

a. Theory

We consider two non-concentric cylinders of radii a and b limiting the hollow plasma placed in vacuum (figure 3.5). The distance between the axis of the two cylinders is $0_a 0_b = K$. As in section 3.1, we describe the plasma by its equivalent permittivity and use the quasistatic approximation which is fully justified in this case.

The incident field is a homogeneous plane wave with its magnetic induction $\mathbf{B_i}$ parallel to the cylinder axis and its electric field strength $\mathbf{E_i}$

in the cross-section of the cylinders. We will therefore consider an exciting E_i field of unit amplitude in the whole space; because of the quasi-static approximation it represents quite satisfactorily the incoming E_i field of the plane wave in the neighbourhood of the column where the boundary conditions have to be stated. We note γ the angle between $0_a 0_b = K$ and E_i.

We consider a biaxial coordinate system (see page 1210 of reference 9) characterized by ξ, ψ, z ($-\infty \leq \xi \leq +\infty$ and $0 \leq \psi < 2\pi$). If 2δ is the distance between the points $\xi = +\infty$ and $\xi = -\infty$, and a and b are the

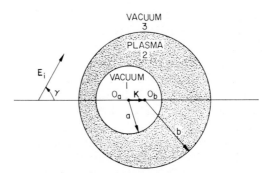

Figure 3.5   Asymmetrical hollow plasma in vacuum

radii of the cylinders (see figure 3.6), ξ_a and ξ_b are the coordinates of those two cylinders when

$$\sinh \xi_a = \frac{\delta}{a} \tag{3-21}$$

$$\sinh \xi_b = \frac{\delta}{b} \tag{3-22}$$

$$\coth \xi_b - \coth \xi_a = \frac{K}{\delta} \tag{3-23}$$

A point on one of the cylinders ξ_a or ξ_b for a given z, is then characterized by the angular coordinate ψ. The Laplace equation separates in this coordinate system:

$$\left(\frac{\cosh \xi + \cos \psi}{\delta^2}\right)^2 \left(\frac{\partial^2}{\partial \xi^2} + \frac{\partial^2}{\partial \psi^2}\right) \phi = 0 \tag{3-24}$$

The general solution by separation of variables is given by

$$\phi = (a_0 + b_0 \xi) + \sum_{n=1}^{\infty} [s_n e^{+n\xi} + t_n e^{-n\xi}](u_n \sin n\psi + v_n \cos n\psi) \tag{3-25}$$

Asymmetry Effects in a Hollow Cylindrical Plasma

where n is an integer in order to have a single-valued function. The constant a_0 can be disregarded as it gives no contribution to the fields. The fact that ϕ must be bounded in region 1 leads to the following expression

$$\phi_1 = \sum_{n=1}^{\infty} [A_n e^{-n\xi} \sin n\psi + A_n' e^{-n\xi} \cos n\psi] \qquad (3\text{-}26)$$

It is easily seen that the $n = 0$ term vanishes in all regions because of the boundedness of the field at $\xi = \infty$ and of the boundary conditions between

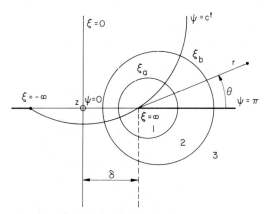

Figure 3.6 Biaxial coordinates

the different regions. In region 2 we have

$$\phi_2 = \sum_{n=1}^{\infty} \{[B_n e^{+n\xi} + C_n e^{-n\xi}] \sin n\psi + [B_n' e^{+n\xi} + C_n' e^{-n\xi}] \cos n\psi\} \qquad (3\text{-}27)$$

In region 3 one has the incoming field represented by ϕ_i and the scattered field represented by ϕ_s. The constant exciting field $\mathbf{E}_i$ is derived from ϕ_i which, in the region $\xi > 0$, has the following form

$$\phi_i = 2\delta \sum_{n=1}^{\infty} (-1)^{n+1} e^{-n\xi} \cos(n\psi + \gamma) \qquad (3\text{-}28)$$

after dropping $\delta \cos \gamma$ which is the $n = 0$ term. The ϕ_s characterizing the scattered field must vanish at infinity ($\xi = 0$, $\psi = \pi$) and takes the form

$$\phi_s = \sum_{n=1}^{\infty} (G_n \cosh n\xi \sin n\psi + G_n' \sinh n\xi \cos n\psi) \qquad (3\text{-}29)$$

A mathematical difficulty arises as ϕ_s of equation (3-29) is infinite at finite distance at the point $\xi = -\infty$. This problem is carefully investigated in a mathematical note at the end of subsection b of this section. It appears that equation (3-29) represents ϕ_s correctly in the neighbourhood of the cylinders when the eccentricity is small; as the boundary conditions are stated on the outer cylinder ξ_b, the procedure is fully satisfactory. It must furthermore be stated that the final formulas (3-34) and (3-35) or (3-45) and (3-46) obtained for G_n and G_n' are rigorously correct if a mirror hollow plasma ($\xi = -\xi_a$; $\xi = -\xi_b$) 'centred' at $\xi = -\infty$ is added to the first one. It is readily seen that the equations are then the same but that the point $\xi = -\infty$ is now excluded from region 3; there are therefore no infinities, the problem is straightforward and leads to the same solution.

Problem: Prove that the above statement concerning a hollow plasma around $\xi = \infty$ and a mirror hollow plasma around $\xi = -\infty$ is correct.

The boundary conditions at the surfaces ξ_a and ξ_b are given by the continuity of the tangential components of the electric field strength and by the continuity of the normal components of the electric displacement.

$$A_n e^{-n\xi_a} = B_n e^{n\xi_a} + C_n e^{-n\xi_a} \tag{3-30}$$

$$B_n e^{n\xi_b} + C_n e^{-n\xi_b} = G_n \cosh n\xi_b - 2\delta(-1)^{n+1} e^{-n\xi_b} \sin \gamma \tag{3-31}$$

$$-\epsilon_0 A_n e^{-n\xi_a} = \epsilon_p B_n e^{n\xi_a} - \epsilon_p C_n e^{-n\xi_a} \tag{3-32}$$

$$\epsilon_p B_n e^{n\xi_b} - \epsilon_p C_n e^{-n\xi_b} = \epsilon_0 G_n \sinh n\xi_b + 2\delta\epsilon_0(-1)^{n+1} e^{-n\xi_b} \sin \gamma \tag{3-33}$$

The above equations are valid for the sin $n\psi$ components and similar equations with primed letters are obtained for the cos $n\psi$ components. The amplitudes of the scattered fields, G_n and G_n' are obtained by solving the two systems of equations (3-30)–(3-33) and (3-30')–(3-33')

$$G_n = 2\delta \sin \gamma (-1)^{n+1} e^{-n\xi_b}$$
$$\times \frac{(\epsilon_p{}^2 - \epsilon_0{}^2)}{\epsilon_p{}^2 \cosh n\xi_b - \epsilon_p \epsilon_0 e^{n\xi_b} \left[\dfrac{e^{2n\xi_b} + e^{2n\xi_a}}{e^{2n\xi_b} - e^{2n\xi_a}}\right] + \epsilon_0{}^2 \sinh n\xi_b} \tag{3-34}$$

$$G_n' = -2\delta \cos \gamma (-1)^{n+1} e^{-n\xi_b}$$
$$\times \frac{(\epsilon_p{}^2 - \epsilon_0{}^2)}{\epsilon_p{}^2 \sinh n\xi_b - \epsilon_p \epsilon_0 e^{n\xi_b} \left[\dfrac{e^{2n\xi_b} + e^{2n\xi_a}}{e^{2n\xi_b} - e^{2n\xi_a}}\right] + \epsilon_0{}^2 \cosh n\xi_b} \tag{3-35}$$

The two resonance frequencies of the sin $n\psi$ component are given by the roots ϵ_1 and ϵ_2 of the denominator of (3-34), while the two resonance frequencies of the cos $n\psi$ component are given by the roots ϵ_1' and ϵ_2' of the denominator of (3-35).

In order to be able to interpret and discuss the results more easily, we go over to the cylindrical system of coordinates as defined in figure 3.6. We then have

$$\xi + i\psi = \ln \frac{2\delta + re^{i\theta}}{-re^{i\theta}} \tag{3-36}$$

$$\sinh n(\xi + i\psi) = \frac{(-1)^n}{2} \left\{ \left[\frac{2\delta}{re^{i\theta}} + 1 \right]^n - \left[\frac{2\delta}{re^{i\theta}} + 1 \right]^{-n} \right\} \tag{3-37}$$

b. Small eccentricity

Combining equations (3-21), (3-22) and (3-23) we have

$$e^{\xi_b} = \frac{b}{K}\left[1 - \frac{a}{b}e^{\xi_b - \xi_a}\right] \tag{3-38}$$

We now define 'small eccentricity' by $K/b \ll 1$. If in this latter case we combine equations (3-21), (3-22), (3-23) and (3-38), we see that when $K = 0_a 0_b$ tends to zero, ξ_a, ξ_b and δ tend to infinity because e^{ξ_a}, e^{ξ_b}, b/K and δ/b are quantities of the same order. Series (3-28) converges then rapidly for $\xi = \xi_b$ (boundary condition) and is well approximated by taking only the $n = 1$ term, i.e.

$$(\phi_1)_{\xi=\xi_b} \cong 2\delta e^{-\xi_b}(\cos\psi\cos\gamma - \sin\psi\sin\gamma) \tag{3-39}$$

Thus the $n = 1$ term is the only one which is appreciably excited and the following form is obtained for the scattered field

$$\phi_s \cong G_1 \cosh\xi \sin\psi + G_1' \sinh\xi \cos\psi \cong \frac{\delta}{r}(G_1 \sin\theta - G_1' \cos\theta) \tag{3-40}$$

after using $\delta/r \gg 1$ in (3-37).

Mathematical note: The infinite value of ϕ_s at the point $\xi = -\infty$ which is inherent in the particular solutions of the Laplace equation which vanish at infinite distance can be investigated as follows. In cylindrical coordinates we know the particular solutions of this equation which vanish at infinity and remain finite elsewhere except at the origin; these are $r^{-n} e^{-in\theta}$. We can therefore choose an arbitrary cylindrical coordinate system and then express $r^{-n} e^{-in\theta}$ in the biaxial system. For reasons of simplicity, we centre the cylindrical system in $\xi = +\infty$ as indicated in figure 3.6. Using $w = \xi + i\psi$ we then have

$$r^{-1}e^{-i\theta} = \frac{1}{\delta}\left(\frac{1}{\tanh w/2 - 1}\right) = \frac{1}{\delta}\frac{\sinh w}{e^{-\xi}e^{-i\psi} - 1} \tag{3-41}$$

We now investigate (3-41) in the region near the tube which is relevant for the boundary condition. For small eccentricity (see section 3.4b) this region is characterized by $e^{-\xi} \ll 1$ and this leads to

$$r^{-1} e^{-i\theta} \simeq -\frac{1}{\delta} (\sinh \xi \cos \psi + i \cosh \xi \sin \psi) \qquad (3\text{-}42)$$

This shows that $\sinh \xi \cos \psi$ and $\cosh \xi \sin \psi$, which are the appropriate combinations at infinity, have also a correct behaviour in the region $e^{-\xi} \ll 1$. Generally, using (3-37) for $\sinh nw$

$$\sinh nw = \frac{(-1)^n}{2} \left[\left(\frac{2\delta}{re^{i\theta}} + 1 \right)^n - \left(\frac{2\delta}{re^{i\theta}} + 1 \right)^{-n} \right] \qquad (3\text{-}37)$$

we see that in the aforesaid region where $r \ll 2\delta$, $\sinh nw$ is practically a linear combination of terms of the form $r^{-m} e^{-im\theta}$. In other words, we have

$$\sinh nw = \sinh n\xi \cos n\psi + i \cosh n\xi \sin n\psi \simeq \sum_{m=1}^{n} C_m r^{-m} e^{-im\theta} \qquad (3\text{-}43)$$

Relation (3-43) shows thus that $(\sinh n\xi \cos n\psi + i \cosh n\xi \sin n\psi)$ correctly represents the scattered field not only at infinitely but also in the region where the boundary condition with the plasma tube is stated.

c. Comparison between theory and experiments

In order to be able to make a quantitative comparison with experiment we must now take the glass walls into account. This situation is described in figure 3.7.

A difficulty arises in the biaxial coordinate system as the two outer cylinders, characterized by radii b and c, are experimentally centred on the same axis. This difficulty is avoided in the following manner: ξ_a, ξ_b and δ are computed from a, b and K as in section 3.4a, and ξ_c is linked to c through

$$\sinh \xi_c = \frac{\delta}{c} \qquad (3\text{-}44)$$

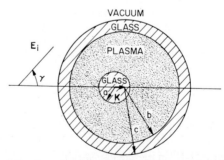

Figure 3.7 Asymmetrical hollow plasma with glass walls

The experimental tube is then exactly described with the exception that the cylinder with radius c is not exactly centred on the cylinder with radius b. The error introduced in this way is small, as the relative difference between b and c is itself small.

Taking the four media of figure 3.7 into account and using exactly the same procedure as in section 3.4a, we finally obtain

$$G_n = -\sin \gamma\, 2\delta(-1)^{n+1} e^{-n\xi_b}$$

$$\times \frac{R_n \epsilon_p^2 - \epsilon_p \epsilon_g (R_n + S_n)\left[\dfrac{e^{2n\xi_b} + e^{2n\xi_a}}{e^{2n\xi_b} - e^{2n\xi_a}}\right] + \epsilon_g^2 S_n}{P_n \epsilon_p^2 - \epsilon_p \epsilon_g (P_n + Q_n)\left[\dfrac{e^{2n\xi_b} + e^{2n\xi_a}}{e^{2n\xi_b} - e^{2n\xi_a}}\right] + \epsilon_g^2 Q_n} \quad (3\text{-}45)$$

$$G_n' = \cos \gamma\, 2\delta(-1)^{n+1} e^{-n\xi_b}$$

$$\times \frac{R_n \epsilon_p^2 - \epsilon_p \epsilon_g (R_n + S_n)\left[\dfrac{e^{2n\xi_b} + e^{2n\xi_a}}{e^{2n\xi_b} - e^{2n\xi_a}}\right] + \epsilon_g^2 S_n}{P_n' \epsilon_p^2 - \epsilon_p \epsilon_g (P_n' + Q_n')\left[\dfrac{e^{2n\xi_b} + e^{2n\xi_a}}{e^{2n\xi_b} - e^{2n\xi_a}}\right] + \epsilon_g^2 Q_n'} \quad (3\text{-}46)$$

where ϵ_g is the permittivity of the glass walls and

$$\begin{matrix} R_n \\ S_n \end{matrix} = \epsilon_0 [e^{2n\xi_b} \mp e^{2n\xi_c}] - \epsilon_g [e^{2n\xi_b} \pm e^{2n\xi_c}]$$

$$\begin{matrix} P_n \\ P_n' \end{matrix} = \epsilon_0 \frac{\sinh n\xi_b}{\cosh n\xi_b}[e^{2n\xi_b} - e^{2n\xi_c}] + \epsilon_g \frac{\cosh n\xi_b}{\sinh n\xi_b}[e^{2n\xi_b} + e^{2n\xi_c}]$$

$$\begin{matrix} Q_n \\ Q_n' \end{matrix} = \epsilon_0 \frac{\sinh n\xi_b}{\cosh n\xi_b}[e^{2n\xi_b} + e^{2n\xi_c}] + \epsilon_g \frac{\cosh n\xi_b}{\sinh n\xi_b}[e^{2n\xi_b} - e^{2n\xi_c}]$$

Again, the two resonant frequencies of the sin $n\psi$ component are given by the roots ϵ_1 and ϵ_2 of the denominator of (3-45) while the two resonant frequencies of the cos $n\psi$ component are given by the roots ϵ_1' and ϵ_2' of the denominator of (3-46). The corresponding computed values of ω_e^2/ω^2 are listed in table 3.3 for $n = 1$.

Table 3.3 Theoretical values ($n = 1$) of the four resonant ω_e^2/ω^2

a (mm)	b (mm)	c (mm)	K (mm)	ξ_a	ξ_b	ξ_c	δ (mm)	$\left(\dfrac{\omega_e^2}{\omega^2}\right)_1$	$\left(\dfrac{\omega_e^2}{\omega^2}\right)_1'$	$\left(\dfrac{\omega_e^2}{\omega^2}\right)_2$	$\left(\dfrac{\omega_e^2}{\omega^2}\right)_2'$
1·45	3·80	5·25	1·0 ± 0·2	2·030	1·160	0·905	5·43	1·98	2·23	8·63	9·25

Examination of table 3.3 shows that the asymmetry introduces a splitting of each of the two resonances which are characteristic of the centred case. This leads to the existence of two resonance doublets. Equation (3-40) shows, that for small eccentricity, each doublet has a $\cos\theta$-component characterized by the amplitude G_1' for the $n = 1$ excitation and a $\sin\theta$-component characterized by the amplitude G_1.

As $K/b \ll 1$ is valid for our discharge tube, we use expression (3-40) for the scattered field ϕ_s. Reflexion experiments are performed; the set-up is thus the one described in section 1.10a and is the same as the one used in section 3.2. One must be careful, however, when comparing the angle θ defined in section 3.2 to the angle θ of the present section. The angle

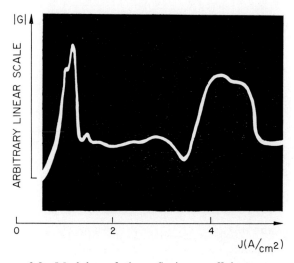

Figure 3.8 Modulus of the reflexion coefficient versus current density for a non-centred hollow cylinder of plasma

discussed in section 3.2 is defined with respect to the incident Poynting vector as there is no preferred direction in the cross-section of the column and reflexion corresponds to $\theta = \pi$. The θ of the present section, however, is defined with respect to the preferred direction of the asymmetrical column, namely $\mathbf{K}$, and the incident electric field $\mathbf{E}_i$ can take an arbitrary direction, characterized by the angle γ, with respect to $\mathbf{K}$. Reflexion corresponds therefore to $\theta = \gamma + \pi/2$. Using equations (3-40), (3-45) and (3-46), we notice that the unprimed component of the reflected electric field strength (not the potential) is proportional to $\sin^2 \gamma$ and that the primed component is proportional to $\cos^2 \gamma$.

Figure 3.8 shows a typical oscillographic record of the amplitude of the reflected electric field strength as a function of the discharge current density $J(\propto \omega_e^2/\omega^2)$. The two resonance doublets appear clearly. The experimental and the theoretical results are compared in table 3.4. Results

Asymmetry Effects in a Hollow Cylindrical Plasma

of three runs are given in order to convey an idea of the dispersion in such experiments.

We have seen that theory predicts, in the case of the reflexion experiment, that each doublet has a $\cos^2 \gamma$ and a $\sin^2 \gamma$ component. This has been tested experimentally by rotating the tube with respect to the measurement set-up. In order that K would not vary during the experiments because of excessive heating of the inner rod (see section 3.2), this was done only for the lower current doublet.

Table 3.4 Comparison between theoretical and experimental results

Theory			Experiments		
			Run 1	Run 2	Run 3
$\dfrac{(\omega_e^2)_1}{(\omega_e^2)_1'}$	0·89	$\dfrac{J_1}{J_1'}$	0·84	0·86	0·87
$\dfrac{(\omega_e^2)_2}{(\omega_e^2)_2'}$	0·93	$\dfrac{J_2}{J_2'}$	0·90	0·90	0·91
$\dfrac{(\omega_e^2)_1}{(\omega_e^2)_2}$	0·23	$\dfrac{J_1}{J_2}$	0·235	0·235	0·24

The oscillographic records of figure 3.9 show that there is striking agreement between theory and experiment. As the detecting system is linear, the amplitudes of the $\sin^2 \gamma$ component for $\gamma = \pi/2$ and $\gamma = \pi/4$ can be compared. Experimentally one finds a ratio of 2:1 which confirms the theoretical value of $(\sin^2 \pi/4)^{-1} = 2$.

d. Conclusions

Theory using a constant equivalent permittivity to describe the plasma shows that lack of centring in a hollow cylindrical plasma leads to a splitting of the two resonances of the centred case, each component of the two resulting doublets having a different angular dependence. The quantitative agreement between theory and experiment is very good. This sustains particularly well the fact that, even for small effects and within the limits described in chapter 1, 'the cold plasma model gives a remarkably accurate description of the common small-amplitude perturbations which are possible for a hot plasma' (see Stix page 6 of reference 3).

It should be stressed that the splitting of a resonance in the case of non-rotational symmetry appears also for an elliptical cross-section of the plasma. It has been shown theoretically[63,25] that an elliptical geometry leads

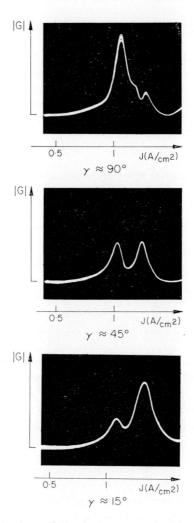

Figure 3.9 Behaviour of the low current doublet as a function of γ

to two cold-plasma resonances, one varying as $\cos^2 \gamma$ and the other as $\sin^2 \gamma$; a qualitative experiment confirmed this.

It is important to be aware of the fact that a small lack of symmetry can introduce very noticeable effects in plasmas. We will study in section 4.3 another instance in which a small asymmetry plays an important role. It will be further seen in section 5.2g that asymmetry can have a significant influence on the temperature spectrum.

Asymmetry Effects in a Hollow Cylindrical Plasma

Problem: Let us introduce elliptical coordinates ξ and η which are related to the ordinary Cartesian coordinates x and y by the defining equations

$$x = c \cosh \xi \cos \eta$$
$$y = c \sinh \xi \sin \eta \qquad \xi > 0; \quad 0 < \eta < 2\pi \qquad (3\text{-}47)$$

Using the suitable combinations of the separable solutions

$$\phi_n = (A_n e^{-n\xi} + B_n e^{n\xi})(C_n \cos n\eta + D_n \sin n\eta) \qquad (3\text{-}48)$$

of Laplace's equation in the plasma and in the vacuum outside, show that the two resonant frequencies of a homogeneous elliptical cylinder of cold plasma are given by

$$\frac{\epsilon_{p1}}{\epsilon_0} = -\coth n\xi_0; \qquad \frac{\epsilon_{p2}}{\epsilon_0} = -\tanh n\xi_0 \qquad (3\text{-}49)$$

where the elliptical contour of the cross-section is defined by $b/a = \tanh \xi_0$ with a and b being the main semi-axes of the ellipse ($a > b$).

CHAPTER 4

Scattering of a Plane Electromagnetic Wave by a Plasma Column in Steady Magnetic Fields (Cold Plasma Approximation)

4.1 Introduction

The interesting and complicated resonance pattern exhibited by a cylindrical plasma in steady magnetic fields was studied experimentally for the first time by Dattner[43]. His was a qualitative work, and Messiaen and Vandenplas[44,45,46,47] performed similar and more detailed experiments on a quantitative basis to obtain results that could be compared with theory.

The scattering of a plane homogeneous wave by a cold plasma column with glass walls is studied theoretically and the results compared to experimental data. The $\mathbf{E}_{inc}$ and $\mathbf{B}_{inc}$ vectors of this incoming wave are

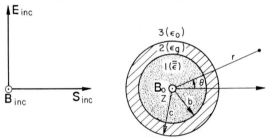

Figure 4.1 Incoming plane wave and plasma cylinder; $\mathbf{B}_0$ is parallel to the axis

indicated in figure 4.1. As $\mathbf{E}_{inc}$ is perpendicular to the axis, the excitation field is said to be TE. The static magnetic induction $\mathbf{B}_0$ is either parallel to the axis of the column or to the Poynting vector $\mathbf{S}_{inc}$ or perpendicular to both the axis and $\mathbf{S}_{inc}$.

Three different problems are investigated. Since the situation is quite intricate, we first give a general outline of these problems together with their main features.

(1) The equilibrium density of the cold plasma cylinder is assumed constant and the anisotropy effects due to the homogeneous steady magnetic fields are studied through the equivalent tensorial permittivity. This permittivity is written in a general form so that the treatment also includes the anisotropic dielectric rod. Because of the intricacies of the equations when $\mathbf{B}_0$ lies in the cross-section of the cylinder, the solution can

Introduction

be obtained only in a Cartesian coordinate system and is then transformed to a cylindrical system so that the boundary conditions can be written down.

When $\mathbf{B}_0$ is parallel to the cylinder axis, the problem can be solved exactly. For moderate values of B_0 the agreement between theory and experiments is very satisfactory. For higher values of B_0 this agreement becomes less satisfactory, but this is not surprising because of the approximations then introduced in the measurement of the plasma electron density. The poorer agreement for higher B_0 is probably partially linked to non-proportionality between the equilibrium electron density and the discharge current in the plasma and also to the greater non-uniformity of the plasma.

When $\mathbf{B}_0$ is perpendicular to the cylinder axis, an approximate solution is obtained which is correct for relatively small magnetic fields as the problem is solved to the second order in β, a dimensionless parameter proportional to B_0. The scattered field is obtained after tedious computations that use the boundary conditions but this field has a polarization different from the incident one. Comparison with experiment shows that the theoretical anisotropy effects are too small to account for all the observed data; even for small values of β there is a considerable influence of the Laplace forces on the equilibrium plasma density distribution. These forces are due to the $\mathbf{B}_0$ in the cross-section and to the drift velocity of the electrons along the axis of the cylinder. (The Laplace force is standard in the French literature for $d\mathbf{F} = I\, d\mathbf{l} \times \mathbf{B} = dq\, \mathbf{v} \times \mathbf{B}$. We prefer to reserve the term Lorentz force for $dq(\mathbf{E} + \mathbf{v} \times \mathbf{B})$). This examination of anisotropy effects is carried out in section 4.2.

(2) In section 4.3 we examine the *asymmetric* inhomogeneity or non-uniformity due to the steady magnetic forces. This supplementary effect will be preponderant in many cases and account for those experimental results which cannot be explained on the basis of section 4.2.

The asymmetrical inhomogeneity arising from the steady magnetic forces is studied assuming that the density is constant in the absence of these forces; no anisotropy is taken into account. The solution of the problem of the asymmetrical inhomogeneity effects is purely TE ($\mathbf{E} \perp$ axis) as no coupling between TE and TM mode ($\mathbf{B} \perp$ axis) is introduced by the non-uniformity. It is characteristic that the different $n\theta$-angular modes ($n = 1$, $n = 2$, $n = 3$, etc.) are coupled through the inhomogeneity. The magnetic field enhances thus the quadrupolar, hexapolar, etc. effects, in agreement with the experimental data. As all the n angular modes are coupled, the chain is broken off in the theoretical computations by neglecting the coefficients of a given order n; this is a satisfactory procedure, as these coefficients tend to zero when n tends to infinity.

The asymmetrical non-uniformity effects, which are much more important than the anisotropy ones, suffice to explain the experimental results obtained in reflexion (TE mode). The experimental and theoretical electron density ratios, corresponding to the first three ($n = 1, 2$ and 3) observable 'cold plasma' resonances, are in very striking agreement.

(3) An effect which is not explained by (1) or (2), however, is the existence of a scattered E_Z even when $B_0 = 0$, the incoming **E** field still being purely transverse as indicated on figure 4.1.

The effect observed consists of a resonance peak of E_Z occurring at the same density for which there exists a resonance of the scattered transverse E field which is well explained by the uniform cold plasma model. The same model is therefore assumed. In this approximation the plasma can be described by a surface charge density Σ due to the HF polarization which gives rise to a surface current density $\mathbf{K} = \Sigma \mathbf{v}_d$ ($\mathbf{v}_d$ is the axial drift velocity of the plasma column). The boundary conditions then lead to the existence of an E_Z which, for $B_0 = 0$, has a sin θ-dependence when the exciting field has a cos θ-dependence. When $B_0 \neq 0$, a more complicated theoretical spectrum is obtained. Both the position and the angular dependence of the resonances are in very good agreement with experimental data. The effect predicted by this model is, however, more than an order of magnitude below that observed in a mercury plasma column.

The present phenomenon can be described as the plasma radio frequency analog of the static field induced by the Roentgen–Eichenwald current. This HF effect due to the axial drift velocity of the plasma is studied in section 4.4.

The investigation of problems (1), (2) and (3) in connexion with the cold magnetized plasma column brings to light two features which are of general interest.

First, it is clear that there are situations where the small asymmetric non-uniformity due to steady magnetic forces leads, in the presence of HF fields, to effects which are more than an order of magnitude greater than those resulting from the anisotropy; this can be of relevance in various plasma set-ups.

Second, the drift velocity $\mathbf{v}_d$ gives rise to an electromagnetic field with **E** parallel to $\mathbf{v}_d$ when an electromagnetic field exists with **E** perpendicular to $\mathbf{v}_d$; this coupling is further complicated and enhanced by the presence of a static magnetic field. Such an effect can be important and should be taken into account in several plasma devices.

Finally, it must be stressed that only the main resonances are explained by the cold plasma approach of the present chapter. Secondary resonances are also observed in magnetic fields, but the latter resonances pertain to non-zero temperature effects which will be studied in chapter 5.

In sections 4.2, 4.3 and 4.4 we will follow papers by Messiaen and Vandenplas[44,45,46,47,50].

4.2 Anisotropy effects

a. General formulation

In a homogeneous and isotropic medium the general solutions of Maxwell's equations can be obtained by the linear combination of two types of solutions. These types, when expressed in a cylindrical coordinate system (r, θ, Z) with no propagation along the Z-axis, will be called the cross-polarized **E**-mode and cross-polarized **H**-mode. These modes are also called TE and TM modes respectively. The cross-polarized **E**-mode (**E** $\perp$ axis) is characterized by H_Z with

$$(\nabla^2 + k^2)H_Z = 0 \quad (4\text{-}1)$$

$$E_r(i/\omega\epsilon r)(\partial H_Z/\partial\theta); \quad E_\theta = (-i/\omega\epsilon)(\partial H_Z/\partial r) \quad (4\text{-}2)$$

The cross-polarized **H**-mode (**H** $\perp$ axis) is characterized by E_Z; H_r and H_θ are obtained from E_Z in a manner analogous to equation (4-2). When the medium is anisotropic or inhomogeneous the two types of solution do not generally remain independent.

In an anisotropic medium described by a tensorial permittivity $\boldsymbol{\epsilon}$ and with an $e^{-i\omega t}$ time dependence, Maxwell's equations take the following form

$$\nabla \times \mathbf{H} = -i\omega\boldsymbol{\epsilon}\mathbf{E} \quad (4\text{-}3)$$

$$\nabla \times \mathbf{E} = i\omega\mu_0\mathbf{H} \quad (4\text{-}4)$$

$$\nabla \cdot \mathbf{H} = 0 \quad (4\text{-}5)$$

$$\nabla \cdot \boldsymbol{\epsilon}\mathbf{E} = 0 \quad (4\text{-}6)$$

where μ_0 is the permeability of vacuum.

The elimination of **E** leads to

$$\nabla \times \boldsymbol{\epsilon}^{-1}\nabla \times \mathbf{H} = \omega^2\mu_0\mathbf{H} \quad (4\text{-}7)$$

A cold plasma in the presence of a static magnetic field can be described by an equivalent tensorial permittivity $\boldsymbol{\epsilon}$ (see section 1.2). The cylindrical coordinate system is the natural one for the present investigations. With no collisions it was found, see equation (1-19), that $\boldsymbol{\epsilon}$ is then given by

$$\boldsymbol{\epsilon} = \epsilon_0 \begin{vmatrix} 1 - \Omega^2 \dfrac{1 - \beta_r^2}{1 - \beta^2} & \Omega^2 \dfrac{\beta_r\beta_\theta + i\beta_Z}{1 - \beta^2} & \Omega^2 \dfrac{\beta_r\beta_Z - i\beta_\theta}{1 - \beta^2} \\ \Omega^2 \dfrac{\beta_r\beta_\theta - i\beta_Z}{1 - \beta^2} & 1 - \Omega^2 \dfrac{1 - \beta_\theta^2}{1 - \beta^2} & \Omega^2 \dfrac{\beta_\theta\beta_Z + i\beta_r}{1 - \beta^2} \\ \Omega^2 \dfrac{\beta_r\beta_Z + i\beta_\theta}{1 - \beta^2} & \Omega^2 \dfrac{\beta_\theta\beta_Z - i\beta_r}{1 - \beta^2} & 1 - \Omega^2 \dfrac{1 - \beta_Z^2}{1 - \beta^2} \end{vmatrix} \quad (4\text{-}8)$$

with $\Omega^2 = (\omega_c/\omega)^2$, $\beta_r = B_{0r}/B_g$, $\beta_\theta = B_{0\theta}/B_g$, $\beta_Z = B_{0Z}/B_g$, and $\beta = B_0/B_g = (\beta_r^2 + \beta_\theta^2 + \beta_Z^2)^{\frac{1}{2}}$, where B_g is the magnetic induction corresponding to a gyrofrequency equal to the frequency of the incoming electromagnetic wave; thus $\beta = eB_0/m\omega$.

When $\boldsymbol{\epsilon}$ is given by equation (4-8), the first of the three scalar equations corresponding to equation (4-7) takes the form

$$\frac{i\Omega^2\beta}{1-\Omega^2}\left[\frac{\sin\theta}{r^2}\frac{\partial^2}{\partial\theta^2}+\frac{\cos\theta}{r^2}\frac{\partial}{\partial\theta}+\frac{\cos\theta}{r}\frac{\partial^2}{\partial r\,\partial\theta}-\frac{\sin\theta}{r}\frac{\partial}{\partial r}\right]H_Z$$
$$+\frac{1}{1-\Omega^2}\left[\frac{\partial}{r^2\,\partial\theta}+\frac{\partial^2}{r\,\partial r\,\partial\theta}\right]H_\theta - \frac{\partial^2 H_r}{(1-\Omega^2)r^2\,\partial\theta^2} = \omega^2\mu_0\epsilon_0 H_r \quad (4\text{-}9)$$

provided that $\mathbf{B_0}$ is perpendicular to the Z-axis, that the terms to second order in β are neglected, and that there is no propagation along the axis.

Equation (4-9) shows that it is well-nigh impossible to solve equation (4-7) directly in cylindrical coordinates when $\mathbf{B_0}$ is perpendicular to the axis. We will show that it is, however, possible to solve equation (4-7) in Cartesian coordinates to second order in β. The solution thus obtained will then be used to generate solutions in cylindrical coordinates, the system in which the boundary conditions are to be stated.

Let us therefore investigate the solutions of equation (4-7) in Cartesian coordinates. The system (x, y, z) is chosen so that the static magnetic induction $\mathbf{B_0}$ is along the z-axis. Please carefully note the difference between the z-axis (parallel to $\mathbf{B_0}$) and the Z-axis which coincides with that of the plasma cylinder. Then

$$\boldsymbol{\epsilon}^{-1} = \frac{1}{\epsilon_0}\begin{vmatrix} \dfrac{K_p - \beta^2}{K_p^2 - \beta^2} & \dfrac{-i\beta\Omega^2}{K_p^2 - \beta^2} & 0 \\ \dfrac{i\beta\Omega^2}{K_p^2 - \beta^2} & \dfrac{K_p - \beta^2}{K_p^2 - \beta^2} & 0 \\ 0 & 0 & \dfrac{1}{K_p} \end{vmatrix} = \begin{vmatrix} M & iK & 0 \\ -iK & M & 0 \\ 0 & 0 & M_3 \end{vmatrix} \quad (4\text{-}10)$$

with $\epsilon_p = \epsilon_0(1 - \Omega^2)$, $K_p = \epsilon_p/\epsilon_0$.

The general form of $\boldsymbol{\epsilon}^{-1}$ as a function of M, M_3 and K is given so as to include the treatment of the anisotropic dielectric rod. Note that $\boldsymbol{\epsilon}^{-1}$ can only be defined when its determinant is different from zero, i.e. when the zero and $\pm\beta$ values of K_p are excluded. The singular value of $\beta = 1$ is also excluded.

The scattering problem we are studying (plane wave with normal incidence on a plasma or dielectric cylinder) is characterized by no propagation

Anisotropy Effects

along the cylinder axis, thus by $\partial/\partial Z = 0$. When performing our transformation from Cartesian to cylindrical coordinates, the Z-axis will always coincide with one of the axes of the Cartesian system. There are two fundamental cases to be considered.

(i) *B_0 parallel to the cylinder axis*

In this case $z \equiv Z$ and $\partial/\partial z = 0$; this leads to

$$\partial H_x/\partial x + \partial H_y/\partial y = 0 \tag{4-11}$$

Thus

$$H_x = \partial \phi/\partial y; \quad H_y = -\partial \phi/\partial x \tag{4-12}$$

Equations (4-7) are then reduced to

$$(\nabla_{Tz}^2 + \omega^2 \mu_0/M_3)\phi = 0 \tag{4-13}$$

$$(\nabla_{Tz}^2 + \omega^2 \mu_0/M)H_z = 0 \tag{4-14}$$

with

$$\nabla_{Tz}^2 \equiv \partial^2/\partial x^2 + \partial^2/\partial y^2.$$

We see that equation (4-13) corresponds to the transverse **H**-mode, called TM mode, and equation (4-14) corresponds to the transverse **E**-mode, called TE mode. These modes are uncoupled as in isotropic media. The problem is straightforward and can be solved exactly. A detailed comparison between theory and experiments will be given in section 4.2b. A rigorous treatment has also been used by Platzman and Ozaki[48] but in quite a different context.

(ii) *B_0 perpendicular to the cylinder axis*

In this case we choose $y \equiv Z$ and we then have $\partial/\partial y = 0$. If $x \equiv Z$ and $\partial/\partial x = 0$ are adopted, isomorphic results are of course obtained. The condition $\partial/\partial y = 0$ leads to

$$H_z = -\partial\phi/\partial x; \quad H_x = \partial\phi/\partial z \tag{4-15}$$

On using (4-15) equations (4-7) reduce to

$$\left[\nabla_{Ty}^2 + \frac{\omega^2 \mu_0}{M}\right]\phi + \frac{iK}{M}\frac{\partial H_y}{\partial z} = 0 \tag{4-16}$$

with $\nabla_{Ty}^2 \equiv (\partial^2/\partial x^2) + (\partial^2/\partial z^2)$

$$\left[\frac{M_3}{M}\frac{\partial^2}{\partial x^2} + \left(1 - \frac{K^2}{M^2}\right)\frac{\partial^2}{\partial z^2} + \frac{\omega^2 \mu_0}{M}\right]H_y + \frac{i\omega^2 \mu_0 K}{M^2}\frac{\partial}{\partial z}\phi = 0 \tag{4-17}$$

The last two equations show that ϕ and H_y are not independent. Eliminating either ϕ or H_y one obtains

$$\left\{\left(\nabla_{Ty}^2 + \frac{\omega^2\mu_0}{M}\right)\left[\frac{M_3}{M}\frac{\partial^2}{\partial x^2} + \left(1 - \frac{K^2}{M^2}\right)\frac{\partial^2}{\partial z^2} + \frac{\omega^2\mu_0}{M}\right]\right.$$
$$\left. + \frac{\omega^2\mu_0 K^2}{M^3}\frac{\partial^2}{\partial z^2}\right\}\binom{\phi}{H_y} = 0 \quad (4\text{-}18)$$

ϕ and H_y are solutions of the same fourth order partial differential equation but as they also satisfy equations (4-16) and (4-17) they are not independent. When

$$(K/M)^2 \ll 1; \quad M/M_3 \approx 1 \quad (4\text{-}19)$$

the fourth order operator can be split into the product of two second order operators and a general solution of equation (4-18) can be obtained in this approximation. This corresponds, for the plasma, to neglecting terms of higher order than the second in β; indeed, one has the following expressions to order β^2

$$K/M \approx -\beta\Omega^2/(1 - \Omega^2) = \gamma \quad (4\text{-}20)$$
$$M/M_3 \approx 1 + \beta^2\Omega^2/(1 - \Omega^2)^2 = 1 + \delta^2 \quad (4\text{-}21)$$

Adopting further the notation

$$\omega^2\mu_0/M_3 = \omega^2\mu_0\epsilon_0 K_P = k_{po}^2 \quad (4\text{-}22)$$

the fourth order operator is equal to the product of the two operators

$$\left\{\frac{\partial^2}{\partial x^2} + \left(1 - \frac{\gamma^2 - \delta^2}{2}\right)\frac{\partial^2}{\partial z^2} + k_{po}^2\left(1 - \frac{\delta^2}{2}\right) + i\gamma k_{po}\frac{\partial}{\partial z}\right\} \quad (4\text{-}23)$$

$$\left\{\frac{\partial^2}{\partial x^2} + \left(1 - \frac{\gamma^2 - \delta^2}{2}\right)\frac{\partial^2}{\partial z^2} + k_{po}^2\left(1 - \frac{\delta^2}{2}\right) - i\gamma k_{po}\frac{\partial}{\partial z}\right\} \quad (4\text{-}24)$$

and this decomposition is valid to order β^2; $\Omega^2 = 1$ and its neighbourhood being excluded.

When $\mathbf{B_0}$ is in the cross-section, resonance effects are damped out before β reaches 0·1 in the existing experimental data[43,44]. We will therefore also neglect β^2 in equations (4-23), (4-24) to simplify the calculations. We then have

$$\left[\nabla_{Ty}^2 + k_{po}^2 + i\gamma k_{po}\frac{\partial}{\partial z}\right]\left[\nabla_{Ty}^2 + k_{po}^2 - i\gamma k_{po}\frac{\partial}{\partial z}\right]\binom{H_y}{\phi} = 0 \quad (4\text{-}25)$$

Each operator of equation (4-25) leads to a type of telegrapher's equation which has a solution of the form

$$aS_{u\pm} = ae^{\mp i\gamma k_{po} z/2}e^{ik(x\cos u + z\sin u)} \quad (4\text{-}26)$$

Anisotropy Effects

where the upper sign corresponds to the operator with the $+$ sign. The two constants of integration are a and u, while

$$k = k_{p0}(1 + (\gamma/2)^2)^{\frac{1}{2}} \approx k_{p0}$$

to order β.

Identification of equation (4-25) with equations (4-16), (4-17)—approximated to first order in β—leads to the four conditions

$$-\partial H_y/\partial z = \mp k_{p0}\partial \phi/\partial z \tag{4-27}$$

$$[\nabla_{Ty}^2 + k_{p0}^2](H_y \mp k_{p0}\phi) = 0 \tag{4-28}$$

or

$$-H_y = \mp k_{p0}\phi \tag{4-29}$$

To each solution H_y of the form

$$(H_y)_u = a_1 S_{u+} + a_2 S_{u-} \tag{4-30}$$

must correspond a solution ϕ of the form

$$(\phi)_u = (1/k_{p0})(a_1 S_{u+} - a_2 S_{u-}) \tag{4-31}$$

Equations (4-30) and (4-31) clearly show that the transverse electric mode and the transverse magnetic mode are linked to each other. When $\beta = 0$, equations (4-27) and (4-28) are no longer fulfilled and the two modes are independent.

b. Theory and experiments with $\mathbf{B}_0$ parallel to the axis

The problem studied is the scattering of a plane homogeneous wave by a cylindrical plasma column when $\mathbf{B}_0$ is parallel to the axis. The z-axis of section 4.2a coincides with the Z-axis of the cylindrical coordinates. The boundary conditions must be stated at the plasma–glass interface and at the glass–vacuum interface (figure 4.1).

The plane wave is characterized by an electric field strength $\mathbf{E}_{\text{inc}}$ and a Poynting vector $\mathbf{S}_{\text{inc}}$. Its $\mathbf{H}_{\text{inc}}$ of unit amplitude is characterized by

$$H_{Z\,\text{inc}} = e^{ik_0 r \cos\theta} = \sum_{n=0}^{\infty} \epsilon_n i^n J_n(k_0 r) \cos n\theta \tag{4-32}$$

where ϵ_n is the Neumann factor. ($\epsilon_n = 1$ for $n = 0$; $\epsilon_n = 2$ for $n > 0$).

In the plasma the equations that must be solved are (4-13) and (4-14), but as the electric field $\mathbf{E}_{\text{inc}}$ of the plane wave is perpendicular to the axis, only the TE mode will be excited and only the solution of equation (4-14) must be sought. In region 1 we thus have for the solution bounded on the axis

$$H_{Z1} = \sum_{n=0}^{\infty} J_n(k_p r)[A'_n \cos n\theta + B'_n \sin n\theta] \tag{4-33}$$

where $k_p{}^2 = \omega^2 \mu_0 / M$. The tangential electric field is

$$E_{\theta 1} = \frac{K}{\omega r} \frac{\partial H_{Z1}}{\partial \theta} - \frac{iM}{\omega} \frac{\partial H_{Z1}}{\partial r} \qquad (4\text{-}34)$$

In region 2 we have

$$H_{Z2} = \sum_{n=0}^{\infty} [M_n' J_n(k_g r) + L_n' N_n(k_g r)] \cos n\theta$$

$$+ \sum_{n=0}^{\infty} [T_n' J_n(k_g r) + U_n' N_n(k_g r)] \sin n\theta \qquad (4\text{-}35)$$

with $k_g = \omega(\epsilon_g \mu_0)^{\frac{1}{2}}$. In region 3 we have $H_{Z3} = H_{Z\text{inc}} + H_{Z\text{sc}}$.

The radial dependence of the scattered field $H_{Z\text{sc}}$ is through a Hankel function of the first kind in order to have the correct asymptotic behaviour when $r \to \infty$; thus

$$H_{Z\text{sc}} = \sum_{n=0}^{\infty} [J_n(k_0 r) + i N_n(k_0 r)][C_n' \cos n\theta + D_n' \sin n\theta] \qquad (4\text{-}36)$$

The boundary conditions for E_θ and H_Z lead to a linear system for the determination of the different coefficients. In this way one obtains transcendental expressions for C_n' and D_n' as a function of Ω^2. For $r = b$ and $r = c$ the arguments of the cylindrical functions, corresponding to a typical experimental set-up, are small so that these functions can be replaced by their asymptotic expressions

$$J_n(kr) \approx \frac{1}{n!} \left(\frac{kr}{2}\right)^n$$

$$N_n(kr) \approx -\frac{(n-1)!}{\pi} \left(\frac{2}{kr}\right)^n \quad \text{when} \quad n \neq 0 \qquad (4\text{-}37)$$

In this approximation algebraic expressions for C_n' and D_n' as functions of Ω^2 are obtained. To do this conveniently, the following unprimed coefficients are defined as a function of the primed ones when $n \neq 0$

$$\{A_n, B_n, M_n, T_n\} = \frac{1}{n!} \left(\frac{k}{2}\right)^n \{A_n', B_n', M_n', T_n'\}$$

$$\{L_n, U_n\} = -\frac{(n-1)!}{\pi} \left(\frac{2}{k}\right)^n \{L_n', U_n'\}$$

$$\{C_n, D_n\} = -i\left(\frac{2}{k}\right)^n \frac{(n-1)!}{\pi} \{C_n', D_n'\} \qquad (4\text{-}38)$$

$$\mathscr{E}_n = \frac{2}{n!} \left(\frac{ik}{2}\right)^n$$

Anisotropy Effects

$E_{\theta 1}$ given by equation (4-34) and written in this approximation has the form

$$E_{\theta 1} = -\frac{i}{\omega \epsilon_p} \sum_{n=1}^{\infty} n r^{n-1}$$

$$\times \left\{ \left[\frac{1 - (\beta^2/K_p)}{1 - (\beta^2/K_p^2)} A_n - \frac{i\beta\Omega^2}{K_p - (\beta^2/K_p)} B_n \right] \cos n\theta \right.$$

$$\left. + \left[\frac{1 - (\beta^2/K_p)}{1 - (\beta^2/K_p^2)} B_n + \frac{i\beta\Omega^2}{K_p - (\beta^2/K_p)} A_n \right] \sin n\theta \right\} \quad (4\text{-}39)$$

This shows that the $\sin n\theta$ and $\cos n\theta$ terms are not independent because of the non-zero β. This is precisely what is predicted by the one-particle approach. An electron accelerated by the incoming HF electric field $\mathbf{E}_{inc}$ (which has a $\cos n\theta$-angular dependence for E_θ) undergoes also, in the cross-section of the column, a rotation due to the axial $\mathbf{B}_0$ and will therefore not only oscillate along $\mathbf{E}_{inc}$, but also in the direction perpendicular to $\mathbf{E}_{inc}$ and to the axis. This gives rise to the $\sin n\theta$ contribution of $E_{\theta 1}$ and thus of the scattered E_θ.

The boundary conditions have thus to be stated for both the sine and cosine components. For $\cos n\theta$ we obtain ($n \neq 0$)

$$b^{-2n} L_n + M_n = A_n$$

$$c^{-2n} L_n + M_n = c^{-2n} C_n + \mathscr{E}_n$$

$$\frac{1}{\epsilon_g}[-b^{-2n} L_n + M_n] = \frac{1}{\epsilon_p}\left[\frac{-i\beta\Omega^2}{K_p - (\beta^2/K_p)} B_n + \frac{1 - (\beta^2/K_p)}{1 - (\beta^2/K_p^2)}\right]_n$$

$$\frac{1}{\epsilon_g}[-c^{-2n} L_n + M_n] = \frac{1}{\epsilon_0}[-c^{-2n} C_n + \mathscr{E}_n] \quad (4\text{-}40)$$

Four analogous equations are obtained for the $\sin n\theta$ components, but then the exciting field given by $\mathscr{E}_n$ is, of course, absent. The system of eight equations is solved and the coefficients of the scattered field are obtained.

$$C_n = -\mathscr{E}_n c^{2n} \frac{[\epsilon_p R_n + \epsilon_g S_n V][\epsilon_p P_n + \epsilon_g Q_n V] - Q_n S_n [\epsilon_g \beta \Omega^2 W]^2}{[\epsilon_p P_n + \epsilon_g Q_n V]^2 - Q_n^2 [\epsilon_g \beta \Omega^2 W]^2} \quad (4\text{-}41)$$

$$D_n = -i\beta \mathscr{E}_n \frac{8 b^{-2n} \epsilon_g^2 \epsilon_p \epsilon_0 \Omega^2 W}{[\epsilon_p P_n + \epsilon_g Q_n V]^2 - Q_n^2 [\epsilon_g \beta \Omega^2 W]^2} \quad (4\text{-}42)$$

where

$$V = (K_p - \beta^2)/[K_p - (\beta^2/K_p)]$$

$$W = 1/[K_p - (\beta^2/K_p)]$$

$$\begin{matrix} P_n \\ Q_n \end{matrix} = \epsilon_0 (b^{-2n} \mp c^{-2n}) + \epsilon_g (b^{-2n} \pm c^{-2n})$$

$$\begin{matrix} R_n \\ S_n \end{matrix} = \epsilon_0 (b^{-2n} \mp c^{-2n}) - \epsilon_g (b^{-2n} \pm c^{-2n})$$

It is easy to show that when $|kr|^2 \ll 1$ [36], the contributions to the scattered field coming from the $n = 0$ term are zero, i.e. the plasma system is transparent for this azimuthal mode in this approximation. The two resonance frequencies of an nth mode are given by the zeros of the denominator of equation (4-41) or equation (4-42) and take the form

$$\frac{\Omega_{1n}^2}{\Omega_{2n}^2} = \left(1 + K_g \left|\frac{Q_n}{P_n}\right|\right)(1 \pm \beta); \qquad K_g = \varepsilon_g/\varepsilon_0 \qquad (4\text{-}43)$$

It is seen that the axial magnetic field gives rise to two main (or cold plasma) resonances by removing the degeneracy that exists between the $e^{in\theta}$ and the $e^{-in\theta}$ mode of the main resonance of the column when $B_0 = 0$.

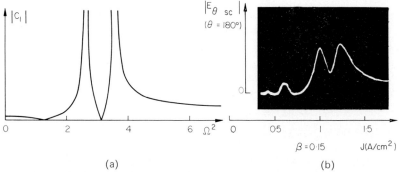

(a)   (b)

Figure 4.2(a) Theoretical curve from equation (4-41) of $|C_1|$ versus $\Omega^2 = \omega_e^2/\omega^2$ when $\mathbf{B_0} \parallel$ axis and $\beta = 0.15$. (b) oscillogram giving the amplitude of the reflected field, $|E_{\theta sc}|$ at $\theta = 180°$, versus the discharge current density $J \propto \omega_e^2$ for $\omega/2\pi = 2.7$ GHz

These resonance frequencies reduce to those obtained by Åström[25] when using a quasi-static approach and neglecting the effect of the glass wall. We see that although the excitation is in $\cos n\theta$, the $\sin n\theta$ terms are excited through the magnetic field as $D_n \propto \beta$. Only the $n = 1$ terms undergo an appreciable excitation in the experiments described, and we take only C_1 and D_1 into account for comparison with observational data.

Problem: Show that in the absence of glass (i.e. uniform plasma cylinder placed in vacuum), equation (4-43) reduces to

$$\frac{\Omega_{1n}^2}{\Omega_{2n}^2} = 2(1 \pm \beta)$$

This last expression clearly shows how the degenaracy of the classical resonance $\omega = \omega_e/\sqrt{2}$ is lifted by the axial magnetic field.

Anisotropy Effects

We first consider the reflected field, i.e. $E_{\theta sc}$ at 180°. The experimental set-up is described in chapter 1. A typical oscillogram (figure 4.2b) giving this reflected field versus the plasma discharge current density is compared to the theoretical curve (figure 4.2a) for a given value of β. The two subsidiary peaks to the left of figure 4.2b are due to temperature effects (see chapter 5). The average discharge current density $J(\propto \omega_e^2)$ is that of the positive column of a Hg low pressure discharge tube with a neutral pressure of 2×10^{-3} Torr ($b = 3\cdot8$ mm; $c = 5\cdot27$ mm). The experimental and theoretical ratios corresponding to the two main peaks are given in table 4.1 [44,46] and show very good quantitative agreement only for rather small values of $B_0(<150$ G, as $\beta = 1$ corresponds to $B \approx 1000$ G).

One must be cautious when making the comparisons in table 4.1, as the connexion between Ω^2 and J is a function of $\mathbf{B_0}$. If we assume that

$$\omega_e^2 = \langle N \rangle e^2/\epsilon_0 m_e = K(\mathbf{B_0})Je^2/\epsilon_0 m_e \tag{4-44}$$

where m_e is the mass of the electron and $\langle N_e \rangle$ the average equilibrium electron density, the comparisons of table 4.1 are possible as each experimental ratio is taken for a given value of B_0. In fact, one expects that K would also be a function of J. When $B_0 = 0$, K is roughly independent of J, but as B_0 has a strong influence on the radial density distribution[51], it is certainly not surprising that, for a given B_0, non-proportionality effects between $\langle N_e \rangle$ and J occur and that they become larger as B_0 becomes greater. This fact has been checked using the dielectric high-frequency resonance probe described in chapter 6. This non-proportionality has been observed and accounts at least partially for the fact that a discrepancy between observational and calculated data appears ($\beta > 0\cdot15$) for rather high values of B_0 ($\beta = 0\cdot15$ corresponds to $B_0 \approx 150$G in the experiments described.) Results of other workers[52], using a lower ω and letting it vary, also show very good agreement between theory and experiments for B_0 up to 50G.

Let us now consider $E_{\theta sc}$ at 90°. To measure it, a waveguide acting as probe must be placed at $\theta = 90°$ but a difficulty of interpretation arises, as the measured field is a combination of the incoming and scattered fields.

Table 4.1 Comparison between theoretical and experimental ratios corresponding to the two resonances ($n = 1$) of a cylindrical plasma when $\mathbf{B_0} \parallel$ axis

	β	0·05	0·1	0·15	0·2	0·3	0·4	0·5
Ω_2^2 / Ω_1^2	observed = J_2/J_1	0·91	0·86	0·81	0·78	0·72	0·65	0·48
	calculated	0·90	0·82	0·74	0·67	0·54	0·43	0·33

Note that the scattered field at $\theta = 90°$ is proportional to D_1 and thus to β; the rise of the two resonances, initially practically degenerate, above the zero level with increasing β is very noticeable in the experiments (see reference 46).

We may therefore conclude that the theory of the cold uniform plasma gives an adequate description of the main scattering properties of a cylindrical plasma column placed in moderate axial magnetic field $\mathbf{B}_0$.

c. Theory with $\mathbf{B}_0$ perpendicular to the axis

When $\mathbf{B}_0$ is perpendicular to the axis of the column, the y-axis of section 4.2a coincides with the Z-axis of the cylindrical coordinates. Equations

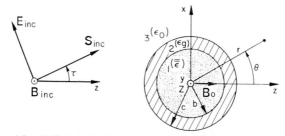

Figure 4.3 Cylindrical plasma column and incoming field when $\mathbf{B}_{0\perp}$ axis. Note that the definition of θ here differs from the definition in section 4.2b

(4-25) are then not separable when expressed in cylindrical coordinates. Following the method used by Whittaker, it is nevertheless possible to construct from the particular solutions S_u, expressed in Cartesian coordinates by equation (4-26), a class of particular solutions in cylindrical coordinates. The problem at hand is characterized by inhomogeneous Dirichlet–Neumann boundary conditions. Considering the elliptical character of equation (4-25), the problem thus admits a unique solution with the result that if the solution constructed is sufficiently general to satisfy the boundary conditions, it is the solution sought.

The solutions that introduce cylindrical functions bounded on the axis are given by

$$S_{n+} = e^{-(i\gamma k_p o r \cos \theta)/2} \int_{-\pi}^{+\pi} du \, e^{ikr(\cos \theta \sin u + \sin \theta \cos u) \pm inu}$$

$$= e^{-(i\gamma k_p o r \cos \theta)/2} J_n(kr) e^{\pm in\theta} \qquad (4\text{-}45)$$

where S_{n+} correspond to S_{u+}; coordinates x and z are linked to r and θ as indicated in figure 4.3. Likewise, we obtain

$$S_{n-} = e^{+(i\gamma k_p o r \cos \theta)/2} J_n(kr) e^{\pm in\theta} \qquad (4\text{-}46)$$

Anisotropy Effects

As we only need the solutions to the first order in β, we have

$$e^{\mp(i\gamma k_{p0} r \cos\theta)/2} = 1 \mp (i\gamma k_{p0} r \cos\theta)/2 \qquad (4\text{-}47)$$

Using equations (4-30), (4-31) and summing over n, we obtain quite general solutions for ϕ and H_z which will prove suitable for our problem.

$$H_Z = \sum_{n=0}^{\infty} (s_n + it_n \zeta r \cos\theta) J_n(k_{p0} r)(A_n' \cos n\theta + B_n' \sin n\theta) \qquad (4\text{-}47)$$

$$\phi = \sum_{n=0}^{\infty} -(1, k_{p0})(t_n + is_n \zeta r \cos\theta) J_n(k_{p0} r)(A_n' \cos n\theta + B_n' \sin n\theta)$$

with $\qquad (4\text{-}48)$

$$\zeta = \gamma k_{p0}/2; \qquad s_n = a_{1n} + a_{2n}; \qquad t_n = a_{2n} - a_{1n}$$

Equations (4-47) and (4-48) show that in the anisotropic plasma the solutions of Maxwell's equations are no longer a linear combination of particular solutions of TE($\mathbf{E} \perp$ axis) and TM ($\mathbf{H} \perp$ axis) type with a $\cos n\theta$ or a $\sin n\theta$-dependence. These solutions are coupled since $\zeta \neq 0$. Taking the asymptotic expansion for $J_n(k_{p0} r)$ when $(k_{p0} r) \ll 1$ and regrouping the terms in $\sin n\theta$ and $\cos n\theta$, we obtain H_{Z1} and E_{Z1} ($\propto \phi$) as given in table 4.2.

In order to state the boundary conditions we must consider the general TE and TM solutions in media 2 and 3. In section 4.2b we saw that in our experimental device only the $n = 0$ and $n = 1$ components of the incident field were important; we will therefore only retain these terms in $H_{Z\text{inc}}$. However, we must now state the boundary conditions for any n, as the expressions for H_{Z1} and E_{Z1} clearly show that the $n > 1$ modes are excited in the plasma through the coupling introduced by $\zeta \neq 0$. This coupling, of course, tends to zero with increasing n.

We consider a plane incoming wave with a Poynting vector $\mathbf{S}_{\text{inc}}$ at an angle τ with $\mathbf{B}_0$. In this way, we include both $\mathbf{B}_0$ parallel to $\mathbf{S}_{\text{inc}}$ and $\mathbf{B}_0$ perpendicular to $\mathbf{S}_{\text{inc}}$ (see figure 4.3). Table 4.2 gives the expression of the fields in different media at whose interfaces the boundary conditions are the continuity of $H_Z, E_\theta, E_Z, H_\theta$. In the isotropic and homogeneous media 2 and 3, E_θ and H_θ are derived from H_Z and E_Z through equation (4-2) or the corresponding relation for H_θ.

The boundary conditions give rise to the following equations:

(1) Four equations for the TE mode in $\cos n\theta$ ($n \neq 0$)

$$b^{-2n} L_n + M_n = A_n + \frac{i\zeta}{2}\left(\frac{2}{\epsilon_{n-1}} \alpha_{n-1} + b^2 \alpha_{n+1}\right)$$

$$c^{-2n} L_n + M_n = c^{-2n} C_n + \delta_{1n} \mathscr{E}_1 \cos\tau$$

$$\frac{1}{\epsilon_g}(-b^{-2n} L_n + M_n) = \frac{1}{\epsilon_p}\left[A_n + \frac{i\zeta}{2}\left(\frac{n-2}{n}\frac{2}{\epsilon_{n-1}}\alpha_{n-1} + b^2 \alpha_{n+1}\right)\right]$$

$$(1/\epsilon_g)(-c^{-2n} L_n + M_n) = (1/\epsilon_0)[-c^{-2n} C_n + \delta_{1n} \mathscr{E}_1 \cos\tau] \qquad (4\text{-}49)$$

Table 4.2 General expression of the fields in the different media

		TE mode ($E_\perp$ axis)	TM mode ($H_\perp$ axis)
Medium 3 (vacuum)	Incoming field	$H_{Z\mathrm{inc}} \cong \mathscr{E}_0 + \mathscr{E}_1 r\{\cos\theta \cos\tau + \sin\theta \sin\tau\}$	$E_{Z\mathrm{inc}} = 0$
	Scattered field	$H_{Z\mathrm{sc}} = C_0[\ln(k_0 r\gamma'/2) - i\pi/2] + \sum_{n=1}^{\infty} r^{-n}[C_n \cos n\theta + D_n \sin n\theta]$ $\ln \gamma'$ is the Euler constant.	$E_{Z\mathrm{sc}} = F[\ln(k_0 r\gamma'/2) - i\pi/2] + \sum_{n=1}^{\infty} r^{-n}[F_n \cos n\theta + G_n \sin n\theta]$
Medium 2 (glass)		$H_{Z2} = L_0 \ln(k_g r\gamma'/2) + M_0 + \sum_{n=1}^{\infty}[(L_n r^{-n} + M_n r^n)\cos n\theta + (T_n r^{-n} + V_n r^n)\sin n\theta]$	$E_{Z2} = L_0{}^* \ln(k_g r\gamma'/2) + M_0{}^* + \sum_{n=1}^{\infty}[(L_n{}^* r^{-n} + M_n{}^* r^n)\cos n\theta + (T_n{}^* r^{-n} + V_n{}^* r^n)\sin n\theta]$
Medium 1 (anisotropic plasma)		$H_{Z1} = \sum_{n=0}^{\infty} r^n \{ [A_n + (i\zeta/2)((2/\epsilon_{n-1})\alpha_{n-1} + r^2 \alpha_{n+1})]\cos n\theta + [B_n + (i\zeta/2)(\beta_{n-1} + r^2\beta_{n+1})]\sin n\theta \}$ $E_{\theta 1} = -\dfrac{i}{\omega\epsilon_\mathrm{p}}\sum_{n=1}^{\infty} r^{n-1}\{[nA_n + (i\zeta/2)((2/\epsilon_{n-1})(n-2)\alpha_{n-1} + nr^2\alpha_{n+1})]\cos n\theta + [nB_n + (i\zeta/2)((n-2)\beta_{n-1} + nr^2\beta_{n+1})]\sin n\theta\}$	$E_{Z1} = i\left(\dfrac{\mu_0}{\epsilon_\mathrm{p}}\right)^{1/2}\sum_{n=0}^{\infty} r^n \{[\alpha_n + (i\zeta/2)((2/\epsilon_{n-1})A_{n-1} + r^2 A_{n+1})]\cos n\theta + [\beta_n + (i\zeta/2)(B_{n-1} + r^2 B_{n+1})]\sin n\theta\}$ $H_{\theta 1} = \dfrac{-1}{k_\mathrm{p0}}\sum_{n=0}^{\infty} r^{n-1}\{[n\alpha_n + (i\zeta/2)((2n/\epsilon_{n-1})A_{n-1} + (n+2)r^2 A_{n+1})\cos n\theta + [n\beta_n + (i\zeta/2)(nB_{n-1} + (n+2)r^2 B_{n+1})]\sin n\theta\}$

where ϵ_{n-1} is the Neumann factor and $A_{-1} = \alpha_{-1} = \beta_0 = B_0 = 0$.

Anisotropy Effects

where δ_{1n} is the Kronecker symbol. These equations, again, clearly show that the n-angular mode is coupled to the $(n - 1)$ and $(n + 1)$ angular modes.

(2) Four equations for the TE mode in $\sin n\theta$ ($n \neq 0$). These equations are similar to equations (4-49) but introduce the constants of the TE fields in $\sin n\theta$ as indicated in table 4.2.
(3) Four equations for the TM mode in $\cos n\theta$ ($n \neq 0$).
(4) Four equations for the TM mode in $\sin n\theta$ ($n \neq 0$).
(5) Four equations for the TE mode when $n = 0$. Here again the small argument asymptotic values of the Bessel functions are used.
(6) Four equations for the TM mode when $n = 0$. More details on these boundary conditions can be found in reference 46 and more particularly in reference 36.

To obtain the amplitudes C_n, D_n, F_n and G_n characterizing the scattered field we must, because of the coupling between the different n-angular modes, solve an infinite set of equations. An approximate solution can be obtained by interrupting the chain, i.e. by setting the $n + 1$ terms equal to zero. This is a very satisfactory procedure as these terms are ζ times smaller than the relevant n terms. We then have a system of $16n + 8$ equations to determine the $16n + 8$ constants (and in particular the desired C, D, F and G).

Results

When $\mathbf{B}_0$ is parallel to the Poynting vector $\mathbf{S}_{\text{inc}}$, $\tau = 0$ and the $\cos n\theta$ and $\sin n\theta$ terms are uncoupled. As only the $\cos n\theta$ terms are excited by the incoming field (see the expression for $H_{Z\text{inc}}$), no $\sin n\theta$ component will be present in the scattered field. Neglecting the coefficients characterized by $n > 2$, the forty equations corresponding to equations (4-49) and following lead to

$$C_0 = 0 \tag{4-50}$$

$$C_1 = -c^{-2n} \frac{[1 + (b\zeta/2)^2]\chi_1^+ - \chi_1^-(b\zeta/2)^2 2w}{[1 + (b\zeta/2)^2]\Delta_1^+ - \Delta_1^-(b\zeta/2)^2 2w} \mathscr{E}_1 \tag{4-51}$$

where

$$w = 1 + 2\{\ln(\gamma' k_0 b/2) - i\pi/2\}$$

$$\chi_n^{\mp} = \epsilon_p R_n \mp \epsilon_g S_n; \qquad \Delta_n^{\mp} = \epsilon_p P_n \mp \epsilon_g Q_n$$

and ln γ' is the Euler constant.

$$C_2 = \frac{\zeta^2 \epsilon_0 \epsilon_g Q_2 \mathscr{E}_0 / \Delta_2}{Q_2[1 + b^2\zeta^2(1 - \epsilon_g Q_2/2\Delta_2)]} \tag{4-52}$$

$$F_0 = \frac{-4\zeta(\mu_0\epsilon_p)^{\frac{1}{2}} \epsilon_g \mathscr{E}_1}{[1 + (b\zeta/2)^2]\Delta_1^+ - \Delta_1^-(b\zeta/2)^2 2w} \tag{4-53}$$

$$F_1 = \frac{-\zeta^3 \mathscr{E}_0 \mu_0^{\frac{1}{2}}(1 - \epsilon_g Q_2/\Delta_2)}{4\epsilon_p^{\frac{1}{2}}[1 + b^2\zeta^2(1 - \epsilon_g Q_2/2\Delta_2)]} \tag{4-54}$$

$$F_2 = \zeta \mu_0^{\frac{1}{2}} A_3 / 4\epsilon_p^{\frac{1}{2}} b^2 \approx 0 \tag{4-55}$$

F_2 is thus negligible in the approximation used.

When $\mathbf{B}_0$ is perpendicular to $\mathbf{S}_{\text{inc}}$, $\tau = \pi/2$ and the $\cos n\theta$ terms of the scattered field are excited by $\mathscr{E}_0$ while the $\sin n\theta$ terms are excited by $\mathscr{E}_1$. Neglecting again the coefficients characterized by $n > 2$, we obtain

$$C_0 = C_1 = F_0 = D_2 = G_1 = 0 \tag{4-56}$$

$$D_1 = -\mathscr{E}_1 c^{-2n} \chi_1^+ / \Delta_1^+ \tag{4-57}$$

C_2 and F_1 are again given by equations (4-52, 4-54).

When the theoretical results, equations (4-51)–(4-57), are compared to the experimental data obtained with the plasma of a mercury discharge tube, one observes (except for the TM coefficients) a great quantitative discrepancy. The effect of the steady magnetic field is indeed much greater for the experimental phenomena than that predicted by ζ in the theoretical expression of the scattered field. It must be stressed that the above analysis was carried out with the assumption of uniform (and thus symmetric) equilibrium density in the cross-section of the cylinder. The experimental plasma, however, is obtained in the positive column of a discharge tube and even for small $\mathbf{B}_0$ there is an appreciable influence of the Laplace forces on the equilibrium electron density distribution. These Laplace forces are due to the existence of a $\mathbf{B}_0$ in the cross-section and to the drift velocity of the electrons along the axis of the cylinder. They are therefore responsible for the existence in the experimental plasma, of a radial *asymmetrical inhomogeneity* of the equilibrium density. The effect of these Laplace forces will be taken into account in section 4.3 and we will see that their influence is preponderant for the TE coefficients. The detailed comparison of theory with experiments when $\mathbf{B}_0$ lies in the cross-section will therefore be given in this next section.

It is because these Laplace forces do not intervene when $\mathbf{B}_0$ is parallel to the axis that a detailed comparison between theory and experiments can be carried out in section 4.2b. The model of a cold *uniform* plasma is then quite adequate to describe the behaviour of the main resonances.

Magnetically Induced Asymmetry Effects

The results obtained above for the anisotropic uniform plasma, when $\mathbf{B}_0$ is perpendicular to the axis will be useful, of course, to interpret the response, in the TE mode, of a quiescent plasma in which the Laplace forces do not appear.

4.3 Magnetically induced asymmetry effects

It has just been said that when the static magnetic induction $\mathbf{B}_0$ lies in the cross-section of the cylinder, a magnetically induced asymmetrical inhomogeneity exists, caused by the Laplace forces resulting from $\mathbf{B}_0$ and from the drift velocity of the electrons along the axis. (As already stated in section 4.1, the Laplace force is standard in the French literature for $d\mathbf{F} = I\,d\mathbf{l} \times \mathbf{B} = dq\mathbf{v} \times \mathbf{B}$. We prefer to reserve the term Lorentz force for $dq(\mathbf{E} + \mathbf{v} \times \mathbf{B})$).

We carry out the theoretical analysis[45,47] with the assumption that the effect of these Laplace forces is to alter the equilibrium density $\langle N \rangle$ of an otherwise homogeneous plasma. We then have[54,55]

$$\langle N \rangle = \langle N_0 \rangle e^{-\delta x} \tag{4-58}$$

where δ is a parameter depending on the magnitude B_0 ($\delta \approx eB_0 v_d / 2KT_e$ where v_d is the mean drift velocity of the electrons along the Z-axis and where the x-axis is perpendicular to the cylindrical axis and to $\mathbf{B}_0$). The plasma is then described by the permittivity

$$\epsilon(x) = \epsilon_0(1 - \Omega^2 e^{-\delta x}); \quad \Omega^2 = \frac{e^2 \langle N_0 \rangle}{\epsilon_0 m_e \omega^2} \tag{4-59}$$

The position of the plane incoming wave and of the Cartesian and cylindrical coordinate systems with respect to the plasma cylinder are indicated in figure 4.3. When $\epsilon(x)$ given by equation (4-59) is introduced in equation (4-7), we obtain

$$\omega^2 \mu_0 \mathbf{H} = [1/\epsilon(x)] \nabla \times \nabla \times \mathbf{H} - (\nabla \times \mathbf{H}) \times \nabla[1/\epsilon(x)] \tag{4-60}$$

This equation is again almost impossible to solve in cylindrical coordinates without further simplification. Using $\partial/\partial y \equiv \partial/\partial Z = 0$ and $\nabla \cdot \mathbf{H} = 0$, equation (4-60) leads in Cartesian coordinates to the scalar equations

$$\omega^2 \mu_0 H_y = [\epsilon(x)]^{-1} \nabla_{TY}^2 H_y + \frac{\partial[1/\epsilon(x)]}{\partial x} \frac{\partial H_y}{\partial x} = 0 \tag{4-61}$$

$$[\nabla_{TY}^2 + \omega^2 \mu_0 \epsilon(x)] \Phi = 0 \tag{4-62}$$

with

$$H_x = \partial \Phi / \partial z; \quad H_z = -\partial \Phi / \partial x \tag{4-63}$$

Equations (4-61) and (4-62) indicate that the sole magnetic inhomogeneity does not introduce any coupling between TE and TM modes. As only the TE solutions derived from H_y (recall that $y \equiv Z$) are excited by the plane incoming wave, the investigation is limited to the solutions of equation (4-61), i.e.

$$\left\{\nabla_{TY}^2 + \omega^2\mu_0\epsilon(x) - \frac{\epsilon_0\Omega^2\delta e^{-\delta x}}{\epsilon(x)}\frac{\partial}{\partial x}\right\}H_y = 0 \qquad (4\text{-}64)$$

As δ is order β (the definition of β is given below equation (4-8)), we will seek a solution of equation (4-64) to order δ in a typical experimental set-up. Furthermore, inside the plasma cylinder $\delta x \ll 1$ as $\delta b = 13\beta$ with $\beta \approx 10^{-2}$. We are therefore led to solve the simplified equation

$$\left\{\nabla_{TY}^2 + k_{po}^2(1 + \eta x) - \eta\frac{\partial}{\partial x}\right\}H_y = 0 \qquad (4\text{-}65)$$

with, again, $k_{po}^2 = \omega^2\mu_0\epsilon_0(1 - \omega_e^2/\omega^2)$ and the condition that $\eta b = \delta b \Omega^2(1 - \Omega^2)^{-1} \ll 1$, that is, excluding $\Omega^2 = 1$ and its neighbourhood. If we set $H_y = f e^{\eta x/2}$ then f satisfies the equation

$$\{\nabla_{TY}^2 + k_{po}^2(1 + \eta x)\}f = 0 \qquad (4\text{-}66)$$

to first order in η.

Now equation (4-66) can be solved directly in cylindrical coordinates to first order in η by using a perturbation method.

We set

$$f = \sum_{n=-\infty}^{+\infty} u_n J_n(k_{po}r)[1 + \eta v_n(r, \theta)]e^{in\theta} \qquad (4\text{-}67)$$

where $J_n(k_{po}r)$ are the solutions, bounded at the origin, of equation (4-66) when $\eta = 0$. Introducing equation (4-67) in (4-66) and neglecting the terms of order η^2, we obtain the equation for $v_n(r, \theta)$

$$(\nabla_T^2 + k_{po}^2)J_n(k_{po}r)v_n(r, \theta)e^{in\theta}$$
$$= (rk_{po}^2/2i)J_n(k_{po}r)[e^{i(n+1)\theta} - e^{i(n-1)\theta}] \qquad (4\text{-}68)$$

To separate the variables, we set

$$v_n(r, \theta) = \mathscr{A}_{n+}(r)e^{i\theta} - \mathscr{A}_{n-}(r)e^{-i\theta} \qquad (4\text{-}69)$$

This leads to an equation for $\mathscr{A}_{n+}(r)$ and to one for $\mathscr{A}_{n-}(r)$

$$\left\{J_n(k_{po}r)\frac{d^2}{dr^2} + \left[2\frac{d}{dr}J_n(k_{po}r) + \frac{J_n(k_{po}r)}{r}\right]\frac{d}{dr}\right.$$
$$\left. - \frac{1 \pm 2n}{r^2}J_n(k_{po}r)\right\}\mathscr{A}_{n\pm}(r) = \frac{rk_{po}^2}{2i}J_n(k_{po}r) \qquad (4\text{-}70)$$

We seek solutions of equation (4-70) that are bounded on the axis, and as the boundary conditions are stated in the neighbourhood of this axis, we take $J_n(k_{p0}r) \propto r^{|n|}$. One then finds a solution for $\mathscr{A}_{n\pm}$ of the form

$$\mathscr{A}_{n\pm} = \frac{r^3 k_{p0}^2}{2i} [6|n| \mp 2n + 8]^{-1} \qquad (4\text{-}71)$$

The solution to first order in η of equation (4-65), where $H_y \equiv H_{Z1}$, is thus

$$H_{Z1} = e^{(\eta/2)r \sin \theta} \sum_{n=-\infty}^{n=+\infty} u_n J_n(k_{p0}r) \left[1 + \eta \frac{k_{p0}^2 r^3}{4i} \right.$$
$$\left. \times \left(\frac{e^{i\theta}}{3|n| - n + 4} - \frac{e^{-i\theta}}{3|n| + n + 4} \right) \right] e^{ni\theta} \qquad (4\text{-}72)$$

As we have limited $J_n(k_{p0}r)$ to its first term, we see through equation (4-72) that in this approximation the contribution of $v_n(r, \theta)$ to H_Z is nil. The final expression of H_{Z1} is thus

$$H_{Z1} = [1 + (\eta/2)r \sin \theta] \sum_{n=0}^{\infty} r^n (A_n \cos n\theta + B_n \sin n\theta) \qquad (4\text{-}73)$$

The components $E_{\theta 1}$ and E_{r1} of the electric field strength in the plasma are deduced from H_{Z1} through

$$\begin{vmatrix} E_{r1} \\ \\ E_{\theta 1} \end{vmatrix} = -\frac{1}{i\omega\epsilon_p(1 + \eta r \sin \theta)} \begin{vmatrix} \frac{\partial H_{Z1}}{r \partial \theta} \\ \\ -\frac{\partial H_{Z1}}{\partial r} \end{vmatrix} \qquad (4\text{-}74)$$

where

$$\epsilon_p/\epsilon_0 = 1 - e^2 \langle N_0 \rangle / m_e \epsilon_0 \omega^2 = 1 - \Omega^2.$$

We now have to express the continuity of H_Z and E_θ at $r = b$ and $r = c$. We only have TE solutions and the general expression of the solutions outside the plasma is given in table 4.2 (media 2 and 3) of the previous section. In order to state the boundary conditions which involve H_{Z1} and $E_{\theta 1}$ in the plasma, we have to group the terms of order n in equation (4-73) to obtain:

$$H_{Z1} = \sum_0^\infty r^n \{[A_n + (\eta/4)(B_{n+1}r^2 - B_{n-1})] \cos n\theta$$
$$+ [B_n + (\eta/4)(2\epsilon_{n-1}^{-1} A_{n-1} - r^2 A_{n+1})] \sin n\theta\} \qquad (4\text{-}75)$$

where $B_0 = B_{-1} = A_{-1} = 0$ and ϵ_n is the Neumann factor. The interested reader will find the details concerning the boundary conditions in reference 47.

Scattered field

The set of equations resulting from the boundary conditions have to be solved to obtain the amplitudes C_n and D_n of the scattered field. We now make a very important remark concerning the relative influence, when $\mathbf{B}_0$ is perpendicular to the axis, of anisotropy as treated in section 4.2c and of asymmetrical inhomogeneity as treated in the present section. Examination of H_{Z1} in table 4.2, on one hand, and of H_{Z2} of equation (4-75), on the other hand, shows that the $i\zeta/2$ of the anisotropy is replaced by the $\eta/4$ of the asymmetrical inhomogeneity. We have for our experimental conditions

$$\left|\frac{\eta/4}{\zeta/2}\right| = \left|\frac{\delta}{\beta k_{p0}}\right| \approx \left|\frac{65}{(1-\Omega^2)^{\frac{1}{2}}}\right| \tag{4-76}$$

and this ratio varies between 45 and 31 in the domain $3{\cdot}1 \leq \Omega^2 \leq 5{\cdot}3$ which is of interest for the experiments. Now, as the TE solutions are generated in both cases by H_{Z1}, we see immediately why the inhomogeneous effects supersede, because of the orders of magnitude involved, the anisotropy effects in the TE mode. This is the reason why one has to go further down the chain than for the anisotropy. Terms up to order $n = 3$ are retained, while it is assumed that the $n = 4$ terms are negligible.

When $\mathbf{B}_0$ is parallel to $\mathbf{S}_{\text{inc}}$, $\tau = 0$ and a tedious elimination leads to the expression of the coefficients ($C_1, C_2, C_3, D_1, D_2, D_3$) of the scattered field in the surrounding vacuum. Since the expression for these coefficients is as cumbersome as that of the coefficients given by equations (4-53) to (4-57), we will not give these coefficients, but refer the reader to reference 47.

When $\mathbf{B}_0$ is perpendicular to $\mathbf{S}_{\text{inc}}$, τ is equal to 90°. Then $C_1 = D_2 = C_3 = 0$, as they are not excited by the incoming field. Again tedious computations lead to the coefficients (D_1, C_2, D_3) of the scattered field in the surrounding vacuum.

Comparison between theory and experiments

We assume that the anisotropy effects of section 4.2c and the magnetically induced asymmetry effects of the present section can be superposed. This assumption is valid, as both calculations were carried out to first order in the perturbation factor β. As already stressed above, the asymmetrical inhomogeneity effects are much more important than the anisotropy effects for the TE mode.

TE Mode

According to what has just been said, the theoretical expressions for the inhomogeneous plasma will be compared with the experimental data when the TE mode is studied. With the $n = 4$ terms set equal to zero, we obtain for the reflected field when $\mathbf{B_0}$ is parallel to $\mathbf{S}_{\text{inc}}$,

$$E_{\theta\,\text{refl}} = (E_{\theta\,\text{sc}})_{\theta=180°} = -\frac{i}{\omega\epsilon_0 r^2}\left(C_1 - 2\frac{C_2}{r} + 3\frac{C_3}{r^2}\right) \quad (4\text{-}77)$$

where C_1, C_2 and C_3 are those for $\tau = 0$. When $\mathbf{B_0}$ is perpendicular to $\mathbf{S}_{\text{inc}}$, the reflected field is

$$E_{\theta\,\text{refl}} = (E_{\theta\,\text{sc}})_{\theta=270°} = -\frac{i}{\omega\epsilon_0 r^2}\left(D_1 + \frac{2C_2}{r} - \frac{3D_3}{r^2}\right) \quad (4\text{-}78)$$

where D_1, C_2, D_3 are those for $\tau = 90°$. Note carefully that the reflected field $E_{\theta\,\text{refl}}$ corresponds to $\theta = 270°$ in this case (see figure 4.3).

The above coefficients have been calculated as a function of Ω^2 for different values of β and for values of the parameters corresponding to experiments. This leads to the following conclusions.

(1) Each term becomes infinite for several values of Ω^2; and thus gives rise to resonances. These are very near the values corresponding to the multipolar resonances $n = 1$, $n = 2$, $n = 3$ which occur for $\mathbf{B_0} = 0$; it must be remembered that these azimuthal n-modes are degenerate for a plasma column in vacuum but that the degeneracy is lifted by the glass wall. $\mathbf{B_0}$ thus alters slightly the values of Ω^2 for which these resonances take place but its chief effect is to enhance them through the coupling between the modes introduced by the asymmetrical inhomogeneity. If the calculations had not been limited to order $n = 3$, the multi-polar resonances of order $n > 3$ would have, of course, been present.

(2) Outside the immediate vicinity of each resonance, the term C_1 of equation (4-77) and the term D_1 of equation (4-78) are dominant and give, with a good approximation, the behaviour of the scattered field. As C_1 and D_1 differ only from contributions of order η^2 which are small with respect to the main one, this explains why the $\mathbf{B_0} \parallel \mathbf{S}_{\text{inc}}$ case and the $\mathbf{B_0} \perp \mathbf{S}_{\text{inc}}$ case differ so little.

In figure 4.4a and figure 4.4b the theoretical curve and the experimental oscillogram ($\omega/2\pi = 2\cdot7$ GHz) are compared when $\mathbf{B_0} \parallel \mathbf{S}_{\text{inc}}$ and $\delta = 0\cdot15$. In figure 4.5a and figure 4.5b the same comparisons are made when $\mathbf{B_0} \perp \mathbf{S}_{\text{inc}}$. Both experiments and theory show that these two cases differ very little in practice. Note that the subsidiary peaks appearing to the left of the main resonance are due to temperature effects (see chapter 5) and that they are, of course, not included in the present cold plasma theory. The theoretical ratios of the resonance values of Ω^2 are compared in table 4.3 to the ratios of the experimental resonance values of the current density $J(\propto \Omega^2)$. The agreement between theory and experiment is strikingly good. When β becomes greater than $0\cdot01$, the third resonance is too strongly

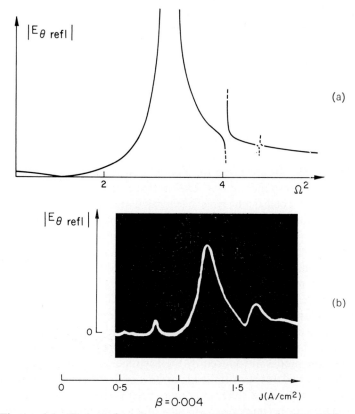

Figure 4.4 Curves for the reflected field $|E_{\theta\,\text{refl}}|$ ($\theta = 180°$, $\mathbf{B}_0 \parallel \mathbf{S}_{\text{inc}}$) as a function of Ω^2 when $\delta = 0.15\ \text{cm}^{-1}$ and $r = 2\ \text{cm}$: (a) theoretical curve; (b) oscillogram where the discharge current density $J \propto \Omega^2$ is the abscissa

damped to be distinguished. This phenomenon is not analysed in detail in this approach with no damping included, but it is nevertheless predicted by the narrowness of the $n = 3$ resonance (see figure 4.4a).

TM Mode

To discuss the influence of the transverse static magnetic field on the TM mode we must consider the anisotropy effects of section 4.2c, as we know (see equation 4-59) that the sole magnetically induced asymmetry does not introduce any coupling between TE and TM modes.

The study of the influence of this anisotropy is not simple because $E_{Z\text{sc}}$ is present even when $\mathbf{B}_0$ is zero. This unexpected fact is due to the

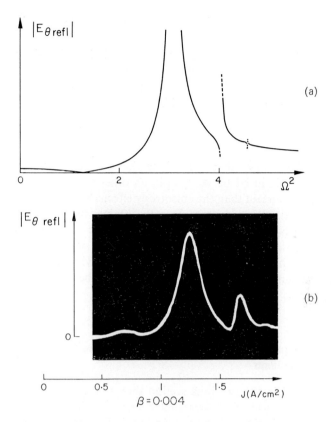

Figure 4.5 Curves for the reflected field $|E_{\theta\text{refl}}|$ ($\theta = 180°$, $\mathbf{B}_0 \perp \mathbf{S}_{\text{inc}}$) as a function of Ω^2 when $\delta = 0.15$ cm^{-1} and $r = 2$ cm: (a) theoretical curve; (b) oscillogram where the current density J is the abscissa

Table 4.3 Comparison between theoretical and experimental ratios corresponding to the first three ($n = 1, 2, 3$) resonances of a cylindrical plasma in reflexion when $\mathbf{B}_0 \perp \mathbf{S}_{\text{inc}}$ or $\mathbf{B}_0 \parallel \mathbf{S}_{\text{inc}}$ and $\delta = 37.5\beta$ cm^{-1}.

	β	0.04	0.08	0.15	0.25	0.35
experiment	$\mathbf{B}_0 \perp \mathbf{S}$	1.37	1.37	1.36	1.34	1.33
$(J)_2/(J)_1$	$\mathbf{B}_0 \parallel \mathbf{S}$	1.37	1.38	1.34	1.32	1.27
theory $(\Omega^2)_2/(\Omega^2)_1$		1.31	1.32	1.32	1.32	1.33
experiment	$\mathbf{B}_0 \perp \mathbf{S}$	*	1.51	1.52	1.51	1.51
$(J)_3/(J)_1$	$\mathbf{B}_0 \parallel \mathbf{S}$	1.48	1.48	1.50	1.48	1.42
theory $(\Omega^2)_3/(\Omega^2)_1$		1.49	1.49	1.49	1.50	1.51

* non-distinct.

coupling between TE and TM modes which is introduced by the axial drift velocity; this latter coupling will be examined in the following section.

For experimental results concerning the TM mode when $\mathbf{B}_0$ is transverse, we refer again to reference 47.

Conclusions

This section shows that the effects due to the small asymmetrical inhomogeneity induced by $\mathbf{B}_0 \perp \mathbf{v}_d$ in an otherwise uniform plasma are much more important than the anisotropy effects. These asymmetry effects suffice to explain the experimental results obtained in reflexion (TE mode).

This suggests that a relatively small asymmetric non-uniformity due to steady magnetic forces can lead, in the presence of HF fields, to effects which are much more important than the more obvious anisotropy effects. This, as was underlined in the introduction, can be of some relevance in various plasma setups.

4.4 High-frequency effect due to axial drift velocity of the plasma column

It was noted in the previous section that a scattered $E_{z\mathrm{sc}}$ component (see figure 4.6) exists, for normal incidence of a plane incoming electromagnetic wave with $\mathbf{E}_{\mathrm{inc}} \perp$ axis, even in the absence of a static magnetic induction $\mathbf{B}_0$.

We will now study a mechanism responsible for the coupling between the cross-polarized E-mode ($\mathbf{E} \perp$ axis) or TE mode, which is the only

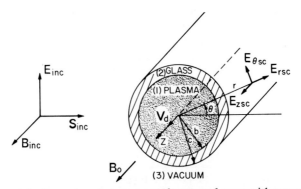

Figure 4.6 Incoming plane wave and scattered wave with respect to the plasma characterized by an axial drift velocity $\mathbf{v}_d$ and in the presence of a possible steady magnetic induction $\mathbf{B}_0$

High-Frequency Effect Due to Axial Drift Velocity

one excited, and the cross-polarized **H**-mode (**H** $\perp$ axis) or TM mode in the absence of $\mathbf{B}_0$ and see how an axial $\mathbf{B}_0$ enhances this coupling.

When $\mathbf{B}_0 = 0$, the effect observed is the existence of a resonance peak of E_{Zsc} at the same electron plasma density $\langle N_e \rangle$ as that for which the main resonance of the TE mode occurs. This last resonance is well explained by cold plasma theory of a uniform column (see e.g. section 3.1 and section 4.2b). We will therefore assume $T_e = 0$ and a constant $\langle N_e \rangle$ in the column as is done throughout this chapter. The new feature is that one takes into account the fact that the plasma possesses an axial drift velocity $\mathbf{v}_d = v_d \mathbf{l}_Z$ ($\mathbf{l}_Z$: unit vector along the Z-axis) as it is generated in the positive column of a low pressure Hg discharge. The plasma can then be described theoretically by a surface charge density of high frequency polarization, which, because of the drift velocity $\mathbf{v}_d$, gives rise to the existence of a scattered E_{Zsc} which, for $\mathbf{B}_0 = 0$, has an angular dependence in $\sin n\theta$ when the excitation by the incoming field is in $\cos n\theta$. This E_{Zsc} displays a resonance for the expected value of $\langle N_e \rangle$. When $\mathbf{B}_0 = B_0 \mathbf{l}_Z \neq 0$, one obtains theoretically a more complicated spectrum which is also in good agreement with the experimental data.

It is very interesting to note that this phenomenon can be considered as the plasma high frequency analog of the static field induced by the Roentgen–Eichenwald current[11,56,57] as is pointed out at the end of this section.

It must be stressed that the agreement between theory and experiment is excellent concerning the position and the angular dependence of the resonances, but that the amplitude of the effect predicted for the cold uniform plasma is at least one order of magnitude below the effect observed with the mercury plasma column.

a. Theory

As in reference 50, we consider a collisionless cold plasma with constant electron equilibrium density $\langle N_e \rangle$. Because of translational symmetry along the Z-axis, the problem is characterized by $\partial/\partial Z \equiv 0$. The linearized hydrodynamical equation for the transport of momentum of the electrons is

$$-i\omega \mathbf{u} + (\mathbf{v}_d \cdot \nabla)\mathbf{u} = -\frac{e}{m}[\mathbf{E} + \mathbf{v}_d \times \mathbf{B} + \mathbf{u} \times \mathbf{B}_{0T}] - \frac{en\mathbf{E}_0}{m\langle N_e \rangle} \quad (4\text{-}79)$$

where $\mathbf{u}$ is the average perturbed electron velocity, $\mathbf{E}$ and $\mathbf{B}$ are the perturbed electric strength and magnetic induction fields in the plasma. The introduction of $\partial/\partial Z \equiv 0$ in equation (4-79) imposes $(\mathbf{v}_d \cdot \nabla)\mathbf{u} = 0$. Furthermore, the term $\mathbf{v}_d \times \mathbf{B}$ can be neglected. Indeed when $\mathbf{v}_d = 0$, **B** is parallel to the Z-axis and this term is exactly zero; when $\mathbf{v}_d \neq 0$ this term is a second order effect as the transverse part of **B** is of order E_{zsc}/c. The term $\mathbf{v}_d \times \mathbf{B}$ is then of order $v_d E_{zsc}/c$ and can be neglected in first approximation when one estimates E_{zsc} (c: velocity of light).

$\mathbf{B}_{0T}$ is the total static magnetic induction resulting from the applied axial $\mathbf{B}_0$ and from the $\mathbf{B}_{0\theta}$ due to the static discharge current density $\mathbf{J}_0$. Neglecting the ionic contribution, we have $\mathbf{J}_0 = -e\langle N_e \rangle \mathbf{v}_d$ and $B_{0\theta} = \mu_0 J_0 r/2$. The effect of this $B_{0\theta}$ field is studied in reference 50 where it is shown that it can be neglected to the order of the approximations used here. We shall therefore only consider the effect of the applied axial $\mathbf{B}_0$. $\mathbf{E}_0 = E_0 \mathbf{1}_Z$ is the axial electrostatic field which is linked to $\mathbf{v}_d$ by the electron mobility μ_e such that $\mathbf{v}_d = \mu_e \mathbf{E}_0$. Equation (4-79) is then written as

$$-e\langle N_e \rangle \mathbf{u} = \boldsymbol{\sigma}\mathbf{E} - \frac{e^2 n v_d}{i\omega m \mu_e} \tag{4-80}$$

with

$$\boldsymbol{\sigma} = \frac{i\epsilon_0 \omega_e^2}{\omega} \begin{bmatrix} \dfrac{1}{1-\beta^2} & -i\dfrac{\beta}{1-\beta^2} & 0 \\ i\dfrac{\beta}{1-\beta^2} & \dfrac{1}{1-\beta^2} & 0 \\ 0 & 0 & 1 \end{bmatrix}$$

where β is as defined after equation (4-8). $\boldsymbol{\sigma}$ is defined as the high frequency tensorial conductivity of the plasma and is independent of $\mathbf{v}_d$. Hence the perturbed plasma current density is

$$\mathbf{J} = -e(\langle N_e \rangle \mathbf{u} + n\mathbf{v}_d) = \boldsymbol{\sigma}\mathbf{E} - en\mathbf{v}_d\left(1 + \frac{e}{i\omega m \mu_e}\right) \tag{4-81}$$

Since $\mu_e \simeq 5 \times 10^3$ m^2 s^{-1} V^{-1} in typical experiments with mercury, $e^2/i\omega m \mu_e \simeq 1\cdot 9 \times 10^{-3}$ (when $\omega/2\pi = 2\cdot 7$ GHz) and can be neglected right away. The plasma is described by Maxwell's equations with sources

$$\text{curl } \mathbf{H} = \mathbf{J} - i\omega\epsilon_0 \mathbf{E} \tag{4-82}$$

$$\text{div } \mathbf{E} = -\frac{en}{\epsilon_0} \tag{4-83}$$

$$\text{curl } \mathbf{E} = i\omega\mu_0 \mathbf{H} \tag{4-84}$$

$$\text{div } \mathbf{H} = 0 \tag{4-85}$$

The perturbed density n can be expressed as a function of $\mathbf{J}$, given by (4-81), through the continuity equation; we obtain

$$-en = \frac{\epsilon_0 \omega_e^2}{\omega^2}\left(\frac{1}{1-\beta^2}\text{ div }\mathbf{E} - \frac{i\beta}{1-\beta^2}\text{ curl}_z\mathbf{E}\right) - \frac{e\text{ div }(n\mathbf{v}_d)}{i\omega} \tag{4-86}$$

One has div $(n\mathbf{v}_d) = n$ div $\mathbf{v}_d + \mathbf{v}_d \cdot$ grad $n = 0$ as $\partial/\partial Z \equiv 0$.

High-Frequency Effect Due to Axial Drift Velocity

Let us first examine this equation in the absence of the static magnetic field ($\beta = 0$). Using equation (4-83) one finds the classic result $n = 0$ in the bulk of the plasma when $\omega \neq \omega_e$ [see e.g. equations (1-26) and (3-20 bis)]. This means that there will be a perturbed surface density and that the surface phenomena will be extremely important in this description.

When $\beta \neq 0$, the result $n = 0$ in the bulk of the plasma no longer holds and we have a perturbed density given by

$$e[1 - \omega_e^2/\omega^2(1 - \beta^2)]n = \frac{i\beta}{1 - \beta^2} \frac{\epsilon_0 \omega_e^2}{\omega^2} \text{curl}_Z \mathbf{E} \tag{4-87}$$

For $\beta \ll 1$ we may neglect this bulk density perturbation with respect to the perturbed surface density in the expression of $\mathbf{J}$ as given by equation (4-81) because the ratio of the bulk drift current to the surface drift current is of the order of $\beta(b/\lambda_0)^2$ ($b/\lambda_0 \ll 1$ in the case treated). We therefore assume that there is no bulk density perturbation.

Problem: Show that above statement on the ratio of the bulk drift current to the surface drift current is correct.

Surface equations

The plasma is connected to the neighbouring glass medium through Maxwell's surface equations

$$\text{curl}_s \mathbf{H} = \mathbf{K} \tag{4-88}$$
$$\text{div}_s \mathbf{D} = \Sigma \tag{4-89}$$
$$\text{curl}_s \mathbf{E} = 0 \tag{4-90}$$
$$\text{div}_s \mathbf{H} = 0 \tag{4-91}$$

and the surface continuity equation

$$-i\omega\Sigma + \text{div}_s \mathbf{J} = 0 \tag{4-92}$$

where Σ is the surface charge density and $\mathbf{K}$ the surface current density. We must now establish the expressions of these two quantities as a function of plasma properties. As the medium 2 surrounding the plasma (medium 1) is a dielectric (see figure 4.6), $J_{2r} \equiv 0$ and equation (4-92) becomes

$$(J_{1r})_{r=b} = [(\boldsymbol{\sigma} \cdot \mathbf{E}_1) \cdot \mathbf{l}_r]_{r=b} = -i\omega\Sigma \tag{4-93}$$

where $\mathbf{l}_r$ is the unit vector along r. Note that equation (4-89) can then be written as

$$(\epsilon_2 E_{2r})_{r=b} = \left\{\left[\left(\epsilon_0 - \frac{\sigma}{i\omega}\right)\mathbf{E}_1\right] \cdot \mathbf{l}_r\right\}_{r=b} = [(\boldsymbol{\epsilon} \cdot \mathbf{E}_1) \cdot \mathbf{l}_r]_{r=b} \tag{4-94}$$

where ϵ is the standard equivalent tensorial permittivity. To obtain **K** we must consider the surface limit of equation (4-81), taking into account that $-en = \Sigma\delta(r-b)$ where δ is the delta function. This leads to $\mathbf{K} = \Sigma\mathbf{v}_d$ and means that the plasma is described as moving as a whole with velocity $\mathbf{v}_d$; the applied high frequency field is responsible for the existence of the charge density Σ and this charge density also undergoes the displacement with velocity $\mathbf{v}_d$. If we had not neglected the bulk density perturbation with respect to the surface perturbation when $\beta \neq 0$, the axial current would have been multiplied by a factor $[1 + 0(\beta b^2/\lambda_0^2)]$. As $\mathbf{v}_d$ only intervenes in the expression of the scattered fields through **K**, this means that the final expressions for G_n and F_n [see equations (4-98) and (4-99)], which are proportional to v_d, would have been multiplied by a factor $[1 + 0(\beta b^2/\lambda_0^2)]$. This last remark fully justifies the procedure followed and the approximation made.

Equation (4-88) then leads to

$$(H_{Z2})_{r=b} = (H_{Z1})_{r=b} \tag{4-95}$$

$$(H_{\theta 2})_{r=b} = (H_{\theta 1})_{r=b} + \left[\frac{(\boldsymbol{\sigma}\mathbf{E}_1)\cdot\mathbf{1}_r}{-i\omega}\right]_{r=b} v_d \tag{4-96}$$

Scattered field

The problem studied is the scattering of a plane homogeneous wave by a cylindrical plasma column with glass wall placed in vacuum (figure 4.6). Equation (4-82) can be written in the plasma as curl $\mathbf{H} = -i\omega\boldsymbol{\epsilon}\cdot\mathbf{E}$ with $\boldsymbol{\epsilon}$ as defined in equation (4-94). The solutions of Maxwell's equations in this medium will therefore have the same form as when $\mathbf{v}_d = 0$. It is only the surface condition (4-96) between plasma and dielectric, where $\mathbf{v}_d$ intervenes, which is different.

It was seen in section 4.2b that the anisotropy does not couple the cross-polarized **H**-mode to the cross-polarized **E**-mode when $\mathbf{B}_0$ is axial. Here, however, the coupling occurs through the boundary condition (4-96), i.e. through v_d. The solutions corresponding to the TE mode and the TM mode in the plasma are deduced from H_{Z1} and E_{Z1}) respectively, which each satisfy a Helmholtz equation. We again use $\boldsymbol{\epsilon}^{-1}$ as given by equation (4-10). The expressions for H_{Z1}, H_{Z2} and H_{Zsc} are, of course, identical to those of section 4.2b and are thus given respectively by equations (4-35), (4-37) and (4-38). E_{Z1}, E_{Z2} and E_{Zsc} have exactly the same analytical expression as the corresponding H_Z. We simply write E_{Zsc} down, as it is the quantity which will be analysed at some length.

$$E_{Zsc} = \sum_{n=0}^{\infty}[J_n(k_0 r) + iN_n(k_0 r)][F_n'\cos n\theta + G_n'\sin n\theta] \tag{4-97}$$

We now state the boundary conditions (4-95) and (4-96) at the two interfaces after having expressed E_r and E_θ as a function of H_Z, and H_θ as a function of E_Z. This leads to a linear system of sixteen equations for each angular mode n and the solution of this system gives the different coefficients A_n', B_n', etc. In this way one finds transcendental expressions for the coefficients C_n', D_n', F_n' and G_n' of H_{Zsc} and E_{Zsc} as a function of Ω^2. For $r = a$ and $r = b$ the arguments of the cylindrical functions corresponding to a typical experimental set-up are small so that these functions can be replaced by their asymptotic expressions as given by formulas (4-37). $J_n(kr)$ is also neglected with respect to $N_n(kr)$ in H_{Zsc} and E_{Zsc}. As has already been stressed, this approximation corresponds to the neglect of the radiation damping. In this way algebraic expressions for the coefficients as functions of Ω^2 are obtained. To do this conveniently, unprimed coefficients are defined as a function of the primed ones; see formulas (4-38).

The formulas obtained for C_n and D_n (TE mode) are, of course, identical to those obtained with $\mathbf{v}_d = 0$; see equations (4-41) and (4-42) of section 4.2b.

For the TM mode, the scattered field E_{Zsc} ($n \neq 0$) is characterized by

$$G_n = 2v_d \mu_0 \epsilon_0 \epsilon_g \mathscr{E}_n$$

$$\times \left\{ \frac{\left[\epsilon_0 K_p P_n + \epsilon_g Q_n \dfrac{K_p - \beta^2}{K_p - \beta^2/K_p}\right] \dfrac{K_p(K_p - 1)}{K_p - \beta^2/K_p} + \beta^2 \epsilon_g Q_n \left[\dfrac{\Omega^2}{K_p - \beta^2/K_p}\right]^2}{\left[\epsilon_0 K_p P_n + \epsilon_g Q_n \dfrac{K_p - \beta^2}{K_p - \beta^2/K_p}\right]^2 - \left[\beta \epsilon_g Q_n \dfrac{\Omega^2}{K_p - \beta^2/K_p}\right]^2} \right\}$$

(4-98)

$$F_n = i 2\beta v_d \mu_0 \epsilon_0 \epsilon_g \mathscr{E}_n \dfrac{\Omega^2}{K_p - \beta^2/K_p}$$

$$\times \left\{ \frac{\epsilon_g Q_n \dfrac{K_p(K_p - 1)}{K_p - \beta^2/K_p} + \left[\epsilon_0 K_p P_n + \epsilon_g Q_n \dfrac{K_p - \beta^2}{K_p - \beta^2/K_p}\right]}{\left[\epsilon_0 K_p P_n + \epsilon_g Q_n \dfrac{K_p - \beta^2}{K_p - \beta^2/K_p}\right]^2 - \left[\beta \epsilon_g Q_n \dfrac{\Omega^2}{K_p - \beta^2/K_p}\right]^2} \right\}$$

(4-99)

with, we recall,

$$K_p = 1 - \Omega^2 \quad \text{and} \quad \dfrac{P_n}{Q_n} = \epsilon_0(b^{-2n} \mp c^{-2n}) + \epsilon_g(b^{-2n} \pm c^{-2n})$$

It can be shown that for $|kr| \ll 1$ the contribution from the $n = 0$ terms is zero.

The physical meaning of (4-98) and (4-99) is clear. For each angular multipole n, the axial drift velocity $\mathbf{v}_d$ couples a scattered cross-polarized **H**-field to the scattered cross-polarized **E**-field excited by the plane incoming wave ($\mathbf{E}_{\text{inc}} \perp$ axis). The two resonances of the TM mode (which degenerate to a single one when $\beta = 0$) are given by the poles of (4-98) and (4-99); these poles are identical to those of the TE mode. The two resonances occur for

$$\Omega_n^2 = \left\{1 + \frac{\epsilon_g}{\epsilon_0} \left|\frac{Q_n}{P_n}\right|\right\}(1 \pm \beta) \tag{4-100}$$

It should be noted that when no steady magnetic field is present ($\beta = 0$), F_n is equal to zero. The picture is then remarkably simple: the incoming cross-polarized **E**-field with its tangential E_θ components in $\cos n\theta$ induces, through $\mathbf{v}_d$, an axial E_{Zsc} with a $\sin n\theta$ dependence.

b. Comparison with experimental data

The axial scattered electric field E_{Zsc} can be easily observed experimentally because it can be separated from the incoming field (the planes of polarization are crossed). Figure 4.7 shows the plasma column irradiated

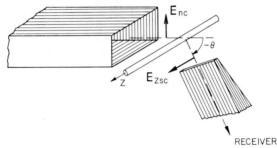

Figure 4.7 Experimental set-up with open waveguide irradiating plasma column and with receiving waveguide

by an open waveguide $\mathbf{E}_{\text{inc}}$ acting as antenna ($\omega/2\pi = 2{\cdot}7$ GHz). The receiving antenna is constituted by a waveguide of decreasing height in order to better localize the field measured, and it can be oriented along different values of θ. The two antennas are carefully oriented (polarization planes exactly perpendicular) so that, in the absence of plasma, the receiving antenna picks up no signal. The remainder of the experimental set-up is as described in chapter 1. Let us recall that the detection is strictly linear and that the oscillographic records give the amplitude of the field measured as a function of the discharge current density $J_0(\propto \omega_e^2)$ in the plasma column. This plasma is obtained in the positive column of a low

High-Frequency Effect Due to Axial Drift Velocity

pressure Hg discharge. The steady magnetic field is created by Helmholtz coils. The $n = 1$ angular mode is the only one appreciably excited and we will consider it only.

Figure 4.8 shows the comparison between experiments (a) and theory (b) for $\theta = 90°$, when $\mathbf{B_0} = 0$*. The high frequency signal on the oscillogram is modulated to 100% and the lower horizontal line corresponds to zero incoming field. It is clear that the coupling between the TE mode and the TM mode is appreciable only near the resonance and that this coupling is nil when $\omega_e \to 0$. The secondary or temperature resonances (see chapter 5) appearing to the left of figure 4.8a are not included in our cold plasma approach, but they are, of course, also coupled by v_d. In exact agreement with theory E_{Zsc} possesses only a $\sin \theta$-dependence when $\beta = 0$. This shows up in figure 4.10a: when $\beta = 0$, E_{Zsc} is practically zero for $\theta = 0°$.

In figure 4.9, experimental data (a) are compared for $\theta = 90°$ to theoretical results (b) when the steady magnetic field is 150 G. The single resonance of $\beta = 0$ splits and the values of J_0, which correspond to the two cold plasma resonances when $\beta \neq 0$, are practically identical to those found for the resonance of $E_{\theta sc}$ (see figure 4.2 and table 4.1).

Figure 4.10 gives the oscillogram of $|E_{Zsc}|$ versus J_0 for $\theta = 0°$ when $\beta = 0$ (a), and when $\beta = 0.04$ (b). This demonstrates conclusively that a component of E_{Zsc} exists, corresponding to F_n of equation (4-99), having a $\cos \theta$-dependence which is proportional to β.

The comparison, as summarized by figures 4.8 to 4.10, shows therefore that there is very good agreement between theory and experiments concerning the position and the angular dependence of the resonances of E_{Zsc}.

Finally, the amplitude of the observed phenomenon is compared to that of the predicted one. This is done for $\beta = 0$. To obtain the amplitude of the resonance theoretically, we consider only the radiation damping which is preponderant in the experimental case considered[36,59], and $J_n(k_0 r)$ must no longer be neglected with respect to $N_n(k_0 r)$ when expressing the boundary conditions. $|G_1|$ remains finite at resonance and is given by

$$|G_1|_{max} = \left| \frac{8\mu_0 v_d \epsilon_g \epsilon_0 \Omega^2 \mathscr{E}_1}{\pi k_0^2 c^2 [\epsilon_p R_1 + \epsilon_g S_1]} \right| \quad (4\text{-}101)$$

with, we recall,

$$\begin{array}{c} R_1 \\ S_1 \end{array} = \epsilon_0(b^{-2} \mp c^{-2}) - \epsilon_g(b^{-2} \pm c^{-2})$$

The maximum theoretical value of $|E_{Zsc}|$ with respect to $|E_{inc}|$ is given in figure 4.8b for $r = 1$ cm. The maximum experimental value of $|E_{Zsc}/E_{inc}|$

---

* The value of v_d is 3×10^5 ms^{-1}. It is calculated from the position of the main resonance in back scattering (see section 3-3). This value agrees with that computed from the determination of the mobility after using Davidson's relation[58].

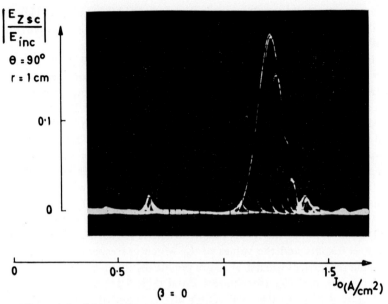

Figure 4.8 (a) $|E_{Zsc}/E_{inc}|$ versus discharge current density $J_0 (\propto \omega_e^2)$ when no steady magnetic field is present ($\beta = 0$)

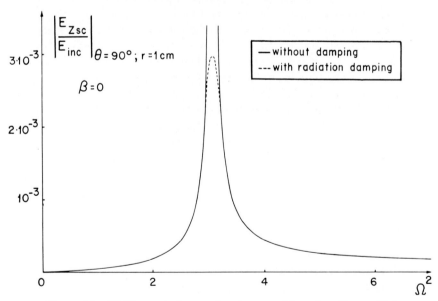

Figure 4.8 (b) Theoretical curve for the same. Note that $\Omega^2 = \omega_e^2/\omega^2$

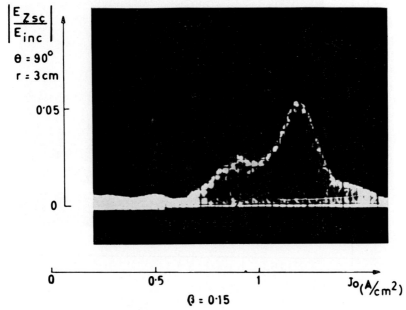

Figure 4.9 (a) Oscillogram giving $|E_{Zsc}/E_{inc}|$ versus discharge current density $J_0 (\propto \omega_e^2)$ for $\theta = 90°$ and $B_0 = 150$ G

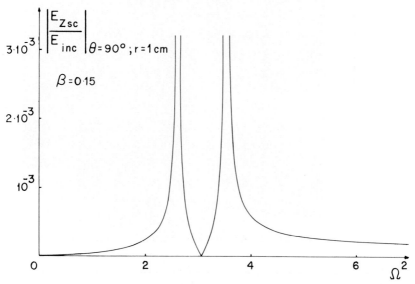

Figure 4.9 (b) Theoretical curve of $|E_{Zsc}/E_{inc}|$ versus Ω^2 for $\theta = 90°$ and $B_0 = 150$ G

125

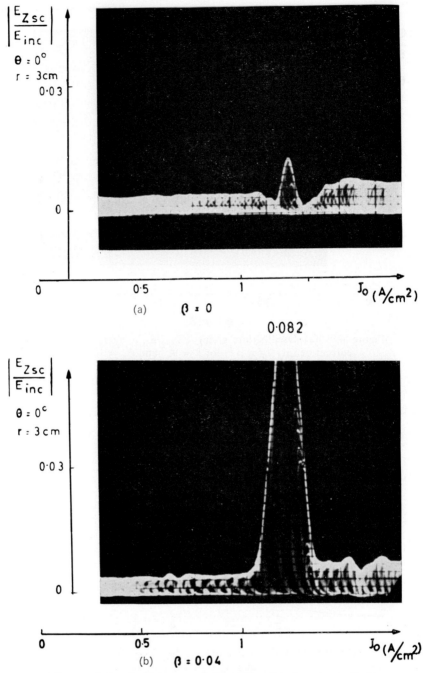

Figure 4.10 Oscillogram giving $|E_{Zsc}/E_{inc}|$ versus discharge current density $J_0(\propto \omega_e^2)$ for $\theta = 0°$ when $\beta = 0$ (a), and $\beta = 0.04$ (b)

High-Frequency Effect Due to Axial Drift Velocity

was obtained through measuring $|E_{\text{inc}}|$ by placing the receiving waveguide at $\theta = 0°$ and parallel to the emitting waveguide. These theoretical and experimental values of $|E_{Z\text{sc}}/E_{\text{inc}}|$ are respectively 0.3×10^{-2} and 16×10^{-2} to 30×10^{-2}. It is seen that the effect observed is at least one order of magnitude higher than the predicted one. This discrepancy does not seem, at first sight, to be explainable by a perturbation in the scattered field due to the presence of the receiving antenna or by the approximations done in the theory. Indeed, when the inhomogeneity and the non-zero temperature T of the plasma is taken into account in simpler cases, the influence on the cold plasma resonance usually remains small (see chapter 5). These two effects should, however, be investigated in this particular instance.

An afterglow experiment shows directly the influence of v_d on the magnitude of $E_{Z\text{sc}}$. The plasma is created by a current pulse in the discharge tube as shown in figure 4.11a. During the ionizing pulse, the electron density rises to a maximum value and the density decreases then as a function of time (time scale: 1 horizontal division corresponds to 10 μs). During the current pulse, the plasma possesses a drift velocity which disappears very rapidly as the current drops, since the collision

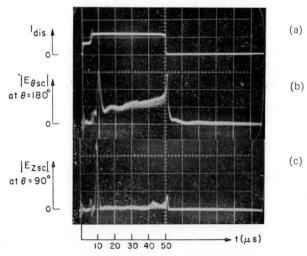

Figure 4.11 (a) Discharge current as a function of time t (1 vertical division = 1 A)
(b) $E_{\theta\text{sc}}$ at $\theta = 180°$ (back-scattered field) versus t (arbitrary linear vertical scale); resonance peak to the left occurs during the electron density rise and peak to the right occurs during the density decay
(c) $E_{Z\text{sc}}$ at $\theta = 90°$ versus t ($E_{\text{inc}} = 25$ vertical divisions); peak to the left occurs when $v_d \neq 0$ and peak to the right occurs when v_d is practically zero ($v_c = 3.5 \times 10^7$ s^{-1})

frequency v_c is approximately 3.5×10^7 s^{-1}. The incident electromagnetic wave is shone on the column as before. Figure 4.11b displays to the left the main or cold plasma resonance in reflexion ($E_{\theta sc}$ for $\theta = 180°$) during the rise of the plasma density, while the resonance peak to the right is obtained during the decay of the density at the time when this density corresponds again to that for the main resonance. The fact that the resonant amplitudes are the same in both cases shows that (except for the drift velocity which does not influence E_θ) the global properties (radial density distribution, etc...) of the plasma do not differ too significantly. Figure 4.11c displays E_{Zsc} as a function of t. One observes, as expected, a resonance peak on the left which occurs at the same time (density) as the E_θ peak; it is characterized by $|E_{Zsc}/E_{inc}| = 0.17$. A second and much smaller resonance peak, occurring again in coincidence with the E_θ peak, is seen to the right; it is characterized by $|E_{Zsc}/E_{inc}| = 0.03$. As demonstrated by figure 4.11b, the essential difference is that $\mathbf{v}_d$ is present in the first instance and that the collisions have considerably reduced it in the second; figure 4.11c shows therefore conclusively that E_{Zsc} is linked to the existence of $\mathbf{v}_d$.

Another experiment which also shows the direct connexion between $\mathbf{v}_d$ and E_{Zsc} is the following: the direction of $\mathbf{v}_d$ is reversed by exactly rotating the tube by an angle of 180° around the vertical. The amplitude and the phase of a reference signal E_{ref} is adjusted in the original position in such a way that the combined signal $E_{ref} + E_{Zsc}$ is equal to zero at resonance when it is displayed as a function of J_0 on an oscilloscope. When $\mathbf{v}_d$ is reversed, one observes $2|E_{ref}|$ at resonance; this shows that the sign of E_{Zsc} is also reversed as predicted by equation (4-99).

c. Final remarks

Attention is drawn to the similarity existing between the phenomenon described here and the classic static experiments of Roentgen and Eichenwald[11,56,57]. In the latter experiments a dielectric is placed in the uniform electrostatic field **E** of a plane condenser, and is displaced parallel to the plates with a velocity **v**. The polarization charges are displaced with this velocity and create a magnetostatic field perpendicular to both **E** and **v**. It is therefore seen that the phenomenon described in this section is, in a sense, the high-frequency plasma analog of the effect due to the static Roentgen–Eichenwald current.

The drift velocity effect is, of course, a direct consequence of the covariance of Maxwell's equations with respect to the Lorentz transformations of special relativity. This explains why the axially moving charges of the whole plasma in the presence of a TE excitation thus give rise to TM fields. In the cold plasma limit, however, the structure of the plasma equations imply, when $\mathbf{B}_0 = B_0 \mathbf{1}_Z$, $\partial/\partial Z \equiv 0$ and $\beta(b/\lambda_0)^2 \ll 1$, the description of the plasma by high frequency surface charges of polarization, and we then obtain the high frequency analog of the Roentgen–Eichenwald current.

It appears that the position of the resonances and the angular distribution of the axial electric field scattered by a mercury plasma are very correctly explained by the theoretical approach which uses a cold uniform plasma column, but that the observed amplitude of the effect is neverthetheless at least an order of magnitude greater than the one predicted by this model. The fact that the drift velocity of a plasma gives rise to a coupling between TE and TM modes and that this coupling is further complicated and enhanced by the presence of a suitably oriented magnetic field, is a feature of general interest. Such an effect could be of some relevance in several plasma set-ups.

CHAPTER 5

Hot Non-Uniform Plasma Column

5.1 Introduction

The present chapter is centred on the theoretical explanation of the temperature or secondary resonance spectrum displayed by a non-uniform plasma column in the absence of a steady magnetic field; see the second part of section 1.10b for detailed experimental information on this. The influence of a steady magnetic field on these temperature resonances will also be considered and, finally, some weak and strong non-linear effects linked to these resonances will be examined.

Since a considerable amount of work has been devoted to these secondary resonances, a short historical introduction appears necessary. As mentioned in section 1.4, early work by Tonks[63] in 1931 indicated that the main resonance frequency of a plasma cylinder of uniform equilibrium density placed in vacuum was equal to $\omega_e/\sqrt{2}$. He observed also two other much smaller resonances which, for fixed ω, occurred at lower densities but he could not account for them in a satisfactory fashion. In connexion with resonant scattering from ionized meteor trails, Herlofson[64] in 1951 made a fundamental study of the problem in the cold plasma limit which included multipolar effects. Among other things he showed that only a single dipolar resonance existed, given by $\omega_e/\sqrt{2}$ for a uniform cylinder. This resonance was damped to an extent that was inversely proportional to the steepness of the density gradient introduced at the edge of the column (the density dropped linearly over a small distance from a constant value to zero at the edge.) Scattering experiments using a discharge tube, made by Romell[92] to compare Herlofson's theoretical results with laboratory data, displayed not only the main resonance but also a series of subsidiary resonances. These additional resonances could not be explained by the cold plasma theory developed by Herlofson. This joint theoretical and experimental work clearly raised and formulated the problem of secondary resonances by indicating that a more general theory was needed to explain them.

Attempts to explain the secondary resonances by an inhomogeneous equivalent permittivity approach proved unsuccessful. Kaiser and Closs[37] found as many resonance frequencies as there were jump discontinuities in a step-like approximation of a generalized Gaussian density distribution. It was shown by Keitel[94] that the field scattered by a plasma cylinder with

Introduction

Gaussian density distribution had a maximum but no marked preferential resonance.

A comprehensive experimental study of secondary resonances in mercury was undertaken by Dattner[43,96,97] in which he varied tube radius and operating frequency, and made an independent determination of plasma density. Further interesting experimental data can be found in the work of many authors[42,44,59,62,79,98,99,100,101,102,103,104]. Experiments with gases other than mercury vapour[59,105,106,107] showed that secondary resonances were, as expected, still present but that the amplitudes of the secondary peaks were, when compared to the amplitude of the main one, significantly smaller than with mercury.

In 1959 Gould[108] analysed the resonance frequencies of a cylinder of constant equilibrium density and *non-zero* temperature placed in vacuum. He found an isolated resonance frequency $\simeq \omega_e/\sqrt{2}$ (the $T = 0$ value being slightly modified by $T \neq 0$) and a discrete resonance spectrum for frequencies higher than ω_e (when ω_e is kept constant and ω is varied). The latter is given by

$$\omega_N^2 = \omega_e^2[1 + 3r_D^2 k_N^2]; \quad N = 1, 2, 3, \ldots$$

where r_D is the Debye radius and k_N is an eigenvalue of

$$k^2 = [\omega^2 - \omega_e^2]/[3KT/m]$$

corresponding to the condition that the perturbed velocity u be set equal to zero at the cylinder boundary. The spacing between the secondary resonance frequencies is, however, generally much too small to account for the experimental results. Note that the above spectrum is very similar to that given by equation (2-30) for the plasma slab-condenser system (section 2.1b). This insufficient spacing is thus inherent in the assumptions made, i.e. constant equilibrium density and $T \neq 0$.

It appeared, therefore, extremely interesting to study a *non-uniform hot* plasma in order to try and explain quantitatively these secondary resonances; this was done in 1960–61 by Vandenplas and Gould[21,60,61]. The plasma is described in a consistent fashion by a perturbed tensorial pressure as indicated in section 1.8. This leads to a sixth-order linear differential system with variable coefficients for the plasma. Unfortunately, it was then not possible to obtain final results that could be compared to experimental data because the calculations proved to be extremely tough for existing computers.

In 1963 a number of investigators simplified the above description of a hot non-uniform plasma by postulating a scalar perturbed pressure $p = \gamma nKT$; the differential system was then reduced to a fourth order one. Non-uniformity in a plasma slab was shown, using various density profiles and several approximations, to increase the spacing between the temperature peaks (Parker[109], Weissglas[110], Hall[111], Gil'denburg[95], Hoh[112]); the case in which the thickness of the boundary layer is much greater than the Debye radius has been investigated by Gil'denburg[113]. At the same time, the solution of the problem was being tackled in cylindrical geometry (Nickel, Parker and Gould[79,80]; Vandenplas and Messiaen[62,114]; Weissglass[115]; Fejer[116]). With a parabolic profile, the computations led to a theoretical representation of the reflected field as a function of the plasma electron density[62] which bore a striking resemblance to the observational data. The position of the resonances is in fair agreement with experiments but not perfect. Very good quantitative agreement is obtained, however, for the position of the main and the first two temperature resonances[79] when a density profile is used which corresponds to the Tonks–Langmuir theory of the positive column in discharge tubes.

A qualitative picture has been given for this phenomenon (Gould[117], Crawford[118]) and it provides a description for the first few secondary resonances. Finally, it should be mentioned that these secondary or temperature resonances are also called Tonks–Dattner resonances by some, following the expression coined by Crawford[118]. Many workers in this field feel, however, that if this appelation does do justice to two important investigators, it does not do justice to three others (namely, Herlofson, Romell and Gould) who, as is shown by the historical sketch above, contributed much to formulate the problem correctly and to elucidate it. As the appelation THRDG resonances would prove rather unwieldy, we shall go on calling them temperature or secondary (subsidiary) resonances, since we feel that this describes their character well.

5.2 Perturbed scalar pressure approximation

a. General formulation in the cylinder case

In order to simplify the equations of section 1.8 describing a hot non-uniform plasma cylinder, one can assume a scalar pressure: $p_{rr} = p_{\theta\theta} = p$ and $p_{r\theta} = 0$. Adding equations (1-109) and (1-110), introducing equation (1-106) one finds

$$p = 2nKT - \frac{\langle N_0 \rangle KT}{i\omega} \frac{dg}{dr} u_r \tag{5-1}$$

If the plasma were uniform, then $p = 2nKT$. Since the gradient becomes important near the edge of the plasma but as the chosen boundary condition is $u_r = 0$ (for more details on this see sections 1.9 and 2.1c), it can be surmised that the last term of (5-1) will not play too important a role.

Perturbed Scalar Pressure Approximation

Anyhow, one sees that postulating $p = \gamma nKT$ is a stronger assumption than the simple scalar hypothesis. For convenience, this stronger assumption has been made by all the workers in the field and γ is usually set equal to 3. In view of what precedes, it might appear that $\gamma = 2$ is a better assumption; one must not forget, however, that the pressure is in fact tensorial and that γ is best treated as a parameter.

The easiest way of establishing a fourth-order differential equation from the equations of section 1.8 when $p = \gamma nKT$ is to start from equations (1-114) and (1-115). We have

$$-i\omega n + \mathbf{u} \cdot \nabla \langle N \rangle + \langle N \rangle \nabla \cdot \mathbf{u} = 0 \tag{5-2}$$

$$-i\omega m \langle N \rangle \mathbf{u} + \gamma KT \nabla n - \frac{KT}{\langle N \rangle} n \nabla \langle N \rangle + e \langle N \rangle \mathbf{E} = 0 \tag{5-3}$$

and

$$\nabla^2 \phi = en/\epsilon_0 \tag{5-4}$$

when the quasistatic approximation $\mathbf{E} = -\text{grad } \phi$ is made. Taking the divergence of equation (5-3), using equations (5-2) and (5-4), we finally obtain the fourth-order equation for ϕ

$$\left(\nabla^2 + \frac{1 - \Omega_0^2 g}{\gamma r_\omega^2} \right) \nabla^2 \phi$$

$$- \frac{1}{\gamma g} \left[\nabla g \cdot \nabla (\nabla^2 \phi) + (\nabla^2 g)(\nabla^2 \phi) - \frac{1}{g} (\nabla g)^2 \nabla^2 \phi + g \frac{\Omega^2}{r_\omega^2} \nabla g \cdot \nabla \phi \right] = 0 \tag{5-5}$$

with $\omega_{e0}^2 = \dfrac{e^2 \langle N_0 \rangle}{\epsilon_0 m_e}$; $\Omega_0^2 = \dfrac{\omega_{e0}^2}{\omega^2}$; $r_\omega^2 = \dfrac{KT}{m_e \omega^2}$; $\langle N \rangle = \langle N_0 \rangle g$

Problem: Derive the above equation.

The solutions of this equation with a parabolic density profile and a Tonks–Langmuir profile will be examined in sections 5.2b and 5.2c respectively. Note that in a uniform plasma equation (5-5) reduces to

$$(\nabla^2 + k^2) \nabla^2 \phi = 0; \quad k^2 = \frac{\omega^2 - \omega_e^2}{\gamma KT/m_e} \tag{5-6}$$

This last equation has been used to solve the problem of the uniform plasma slab-condenser system when $T \neq 0$ (section 2.1c).

Problem: Consider a cylinder of non-uniform hot plasma with an incident plane wave ($\mathbf{E}_i$ perpendicular to the axis and $\mathbf{B}_i$ parallel to the axis, see figure 3.1 or figure 4.1 for example). Using equations (5-2) and (5-3) and the complete Maxwell's equations (no static approximation) show that we have in the plasma

$$\left[\nabla^2 + \frac{\omega^2 - \omega_e^2}{c^2}\right] H_Z = \frac{e}{i\omega m_e}\left[KT\nabla\langle N\rangle \times \nabla\left(\frac{n}{\langle N\rangle}\right) - e\mathbf{E} \times \nabla\langle N\rangle\right]_Z \quad (5\text{-}7)$$

where c is the velocity of light in vacuum, and

$$\left[\nabla^2 + \frac{\omega^2 - \omega_e^2}{\gamma KT/m_e}\right]n = \frac{1}{\gamma}\left[\nabla\left(\frac{n\nabla\langle N\rangle}{\langle N\rangle}\right) - e\nabla\langle N\rangle \cdot \mathbf{E}\right] \quad (5\text{-}8)$$

Let us now consider the physical meaning of these results. When the plasma is uniform, equations (5-7) and (5-8) reduce to

$$\left[\nabla^2 + \frac{\omega^2 - \omega_e^2}{c^2}\right] H_Z = 0 \quad (5\text{-}9)$$

$$\left[\nabla^2 + \frac{\omega^2 - \omega_e^2}{\gamma KT/m_e}\right] n = 0 \quad (5\text{-}10)$$

The transverse electromagnetic waves characterized by H_Z and the longitudinal plasma oscillations characterized by n are uncoupled. Equations (5-7) and (5-8) clearly show that this is no longer true in a non-uniform plasma: the transverse electromagnetic waves and the longitudinal plasma oscillations are coupled by the electron density gradient. In a uniform plasma the comparison of (5-6) with equations (5-9)–(5-10) further illustrates the nature of the quasi-static approximation; the fact that we have the operator ∇^2 instead of $\nabla^2 + (\omega^2 - \omega_e^2)/c^2$ corresponds, of course, to the assumption of an infinite wavelength for the transverse waves since $(\omega^2 - \omega_e^2)/c^2 = k_p^2$ is the square of the propagation vector in the cold plasma considered as equivalent dielectric medium.

b. Parabolic density profile

We summarize in this section the results obtained by Vandenplas and Messiaen[62]. The experimental data is given in section 1.10b. When the incident field is preponderantly dipolar (cos θ-angular dependence in figure 3.1) the temperature peaks have this same angular dependence (Boley[98]). The dipolar excitation was that used in most of the experiments reported above. Equation (5-5) is therefore solved in the dipolar case ($\phi = \Phi(r)\cos\theta$) after adopting the following law for the variation of equilibrium density as a function of radius

$$\langle N\rangle = \langle N_0\rangle g(r) = \langle N_0\rangle\left[1 - \alpha\left(\frac{r}{a}\right)^p\right] \quad (5\text{-}11)$$

Perturbed Scalar Pressure Approximation

As is well known (see reference 120 and later authors), formula (5-11) gives a very satisfactory description of the bulk variation of plasma density if $p = 2$ and $\alpha \simeq 0.6$ when the Debye radius is sufficiently small, but it does not take the sheath effects into account. The Frobenius method applied to equation (5-5) shows that there are two power series solutions which are regular at $r = 0$, namely $f(r)$ with a leading term of order r, and $h(r)$ with a leading term of order r^3.

We wish to obtain the amplitude G of the scattered wave as a function of plasma density for a given incoming plane wave. In order to render the comparison with experiments more direct, the glass wall of the plasma column is introduced. (b: inner radius, c: outer radius, ϵ_g: permittivity of the glass). The continuity of the tangential components of the electric field strength and of the normal components of the electric displacement at $r = b$ and $r = c$ are expressed together with $u = 0$ at $r = b$. More details on this last condition are found in sections 1.9 and 2.1c. A final formula is obtained for the scattered amplitude G^* corresponding to a unit incoming field after some straightforward but somewhat lengthy calculations[121]

$$G = \Delta_G/\Delta \tag{5-12}$$

with

$$c^2\Delta_G = \delta_A\{fK_g[K_g(b^2 - c^2) + b^2 + c^2] - f'[K_g(b^2 + c^2) + b^2 - c^2]\}$$
$$- \delta_B\{hK_g[K_g(b^2 - c^2) + b^2 + c^2] - h'[K_g(b^2 + c^2) + b^2 - c^2]\}$$

$$c^4\Delta = \delta_A\{fK_g[-K_g(b^2 - c^2) + b^2 - c^2] + f'[K_g(b^2 + c^2) - b^2 + c^2]\}$$
$$- \delta_B\{hK_g[-K_g(b^2 - c^2) + b^2 - c^2] + h'[K_g(b^2 + c^2) - b^2 + c^2]\}$$

$$\frac{\delta_A}{\delta_B} = (6 - 8\alpha)\frac{h}{f} + [-\Omega^2(1 - \alpha)^2\rho + 8\alpha - 6]\frac{h'}{f'}$$
$$+ (3 - \alpha)\frac{h''}{f''} + 3(1 - \alpha)\frac{h'''}{f'''}$$

$$\rho = b^2/r_\omega^2 = m_e\omega^2 b^2/KT; \qquad K_g = \epsilon_g/\epsilon_0$$

It is the mean value of the equilibrium density, thus the mean value of ω_e^2 which is proportional to the current I in the discharge column; this value $(\Omega^2)_{Av}$ of $\Omega^2 = \omega_e^2/\omega^2$ is linked to $\Omega_0^2 = \omega_{e0}^2/\omega^2$ by

$$(\Omega^2)_{Av} = \frac{2\Omega_0^2}{b^2}\int_0^b r[1 - \alpha(r/b)^2]\,dr = \Omega_0^2(1 - \alpha/2) \tag{5-13}$$

G is computed as a function of $(\Omega^2)_{Av}$ for values of b and c corresponding

---

* It will be interesting to see in section 5.2e how this formula for G is modified if radiation damping is taken into account.

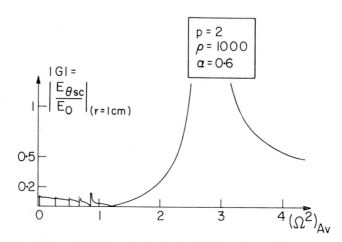

Figure 5.1 (a) Theoretical curve of $|G| = |E_{\theta sc}/E_0|$ ($r = 1$ cm, $\theta = 180°$, $\rho = 1000$) as a function of $(\Omega^2)_{Av} \propto I$ where $E_{\theta sc}$ is the amplitude of the scattered field and E_0 of the incoming one; the peak to right being the main one

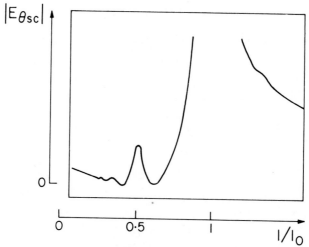

(b) Oscillogram giving $|E_{\theta sc}|$ ($\theta = 180°$) as a function of I/I_0 where I_0 is the discharge current corresponding to the main resonance

to the experimental set-up, namely $b = 3 \cdot 8$ mm, $c = 5 \cdot 27$ mm, $\omega/2\pi = 2 \cdot 7$ GHz. The experimental data shown is obtained with a mercury discharge having an electron temperature of 40,000°K, ρ is then equal to 7500.

The results are computed by representing $h(r)$ and $f(r)$ by their power series expansion in the vicinity of $r = 0$ and then carrying out the integration to $r = b$ by Hamming's method. They are in very close agreement with those obtained in preliminary work where $h(r)$ and $f(r)$ were computed in $r = b$ using the power series expansion of the Frobenius method; this indicates that we can have full confidence in the results listed. It is, however, increasingly difficult to compute $|G|$ when $(\Omega^2)_{Av}$ becomes higher. In fact, the highest value of $(\Omega^2)_{Av}$ for which it is possible to compute a trustworthy value of $|G|$ becomes smaller as ρ or α increases. This is due to increasing difficulties in computing the exact values of h and f at $r = b$; even with the high precision used, h and f and their derivatives become then practically proportional and the two independent solutions cannot be singled out any more. This is the reason why a full computation was run for $\rho = 1000$ even when the experimental value of ρ is as high as 7500. Figure 5.1a gives a clear over-all picture in which the main resonance and five secondary resonances are observed. It is compared to the experimental curve given by figure 5.1b.

For $\rho = 7500$, the highest $(\Omega^2)_{Av}$ for which a trustworthy $|G|$ is observed is $1 \cdot 3$. The first peak on the right of figure 5.2a is the first secondary as a continuity analysis between $\rho = 1000$ and $\rho = 7500$ shows. The main peak is near $(\Omega^2)_{Av} = 3$ for $\rho = 7500$, as we know that for $\rho = \infty$ it is situated at $(\Omega^2)_{Av} \simeq 3 \cdot 1$ (cold plasma). Note that there are theoretically fifteen secondary peaks for $\rho = 7500$. Comparison of this curve with the experimental oscillogram of figure 5.2b, which gives the details of the temperature spectrum, shows a very striking resemblance between theoretical results and observational data. We shall see in section 5.2d how more experimental temperature resonances can be observed by using an appropriate technique. Great care was taken so that the experimental conditions would be identical to the theoretical ones, i.e. an approximately plane incoming wave and rigorous proportionality of the modulus of the reflected field to the quantity represented on the vertical axis of the oscillograms (see section 1.10a). The predicted resonances are given by an infinite value of $|G|$ since no collision effects and no radiation damping (quasi-static approximation) have been included in the theory. The damping effects will be examined in section 5.2e. The fact that the experimental resonance peaks become less and less important as electron density decreases is fully accounted for by the theoretical approach as is clearly seen from figure 5.2a. Comparing figures 5.1a and 5.2a with the cold plasma curve (figure

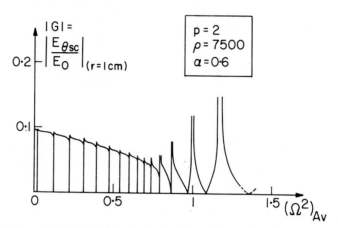

Figure 5.2 (a) Theoretical curve of $|G| = |E_{\theta sc}/E_\theta|$ ($r = 1$ cm, $\theta = 180°$, $\rho = 7500$) as a function of $(\Omega_{Av}^2) \propto I$; the peak to the right being the first secondary

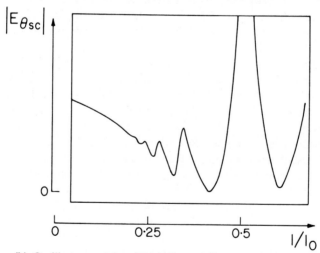

(b) Oscillogram giving $|E_{\theta sc}|$ ($\theta = 180°$) as a function of I/I_0; the peak to the right being the first secondary

Perturbed Scalar Pressure Approximation

3.4a), one sees clearly that the hot plasma scattering behaviour is a sort of superposition of the secondary peaks on an average curve very similar to that of the cold plasma. Of course, the influence of the temperature is also observable on the main peak, but it is small when $\rho = (b^2/r_\omega^2)$ is high. The quantitative comparison between the ratios of resonance discharge currents ($I \propto$ mean equilibrium density) and theoretical ratios of resonance mean equilibrium densities $\propto (\Omega^2)_{Av}$ is listed in table 5.1. The ratios

Table 5.1 Experimental values of I_{n+1}/I_n and corresponding theoretical values of $(\Omega^2)_{Av\ n+1}/(\Omega^2)_{Av\ n}$

	Theory				Experiments
	$\rho = 1000$ $\langle N \rangle = \langle N_0 \rangle$ $\times \left[1 - 0.6\left(\frac{r}{b}\right)^2\right]$	$\rho = 1000$ $\langle N \rangle =$ constant	$\rho = 7500$ $\langle N \rangle = \langle N_0 \rangle$ $\times \left[1 - 0.6\left(\frac{r}{b}\right)^2\right]$	$\rho = 7500$ $\langle N \rangle =$ const.	$\rho = 7500$ Hg
I_1/I_0	0.31	0.33	0.39	0.31	0.52
I_2/I_1	0.79	0.86	0.86	0.98	0.66
I_3/I_2	0.76	0.77	0.87	0.98	0.80
I_4/I_3	0.57	0.56	0.91	0.96	0.86
I_5/I_4	0.078	0.073	0.92	0.95	0.89
I_6/I_5	No more resonances		0.94	0.94	0.915
I_7/I_6			0.94	0.93	0.93
I_8/I_7			0.92	0.91	0.94
I_9/I_8			0.90	0.89	0.92
I_{10}/I_9			0.87	0.87	0.90
I_{11}/I_{10}			0.84	0.84	0.87
I_{12}/I_{11}			0.79	0.77	
I_{13}/I_{12}			0.73	0.72	
I_{14}/I_{13}			0.55	0.56	
I_{15}/I_{14}			0.15	0.14	
				No more resonances	

corresponding to the constant density theory are also listed; the densities of the temperature resonances are given in this latter case by

$$(\Omega^2)_{Av\ n} = \Omega_n^2 = 1 - \frac{3x_n^2}{\rho} \tag{5-14}$$

with $x_n^2 \simeq 26;\ 70;\ 130.5;\ 220;$ etc. Formula (5-14) was established for a homogeneous hot plasma in vacuum, but it can be shown[114] that the glass wall has a negligible effect on the secondary spectrum. We see that it is only in the region for which $\rho \simeq 1000$ that the spacing of the uniform plasma theory is of the order of magnitude of the experimental one for the same value of ρ. For increasing values of ρ it becomes much too small for the first few secondary resonances.

Note that I_{n+1}/I_n is listed in table 5.1 so as to be able to compare successive peaks without carrying along the influence of the position of the first one. In this way one can observe the noteworthy fact that the spacing predicted by the uniform plasma for the first few resonances is much too small to account for the observational data but that beyond I_6/I_5 there is little difference between the uniform and non-uniform plasma predictions. This is traceable to the values taken by h' and f', which generate the electric field, as a function of r. For small values of $(\Omega^2)_{\mathrm{Av}}$, they remain very small and fluctuate little when r varies between 0 and b. There is a smooth transition to higher values of $(\Omega^2)_{\mathrm{Av}}$ for which h' and f' remain small in the core of the plasma but start varying very appreciably towards the edge. Since the density gradient becomes important towards the edge it

Table 5.2 Evolution as a function of α of the theoretical ratio characterizing the first two secondary peaks

	$\rho = 7500$			$\rho = 1000$				
α	0·5	0·6	0·7	0·5	0·6	0·7	0·8	0·9
$\dfrac{(\Omega^2)_{\mathrm{Av}2}}{(\Omega^2)_{\mathrm{Av}1}} = \dfrac{I_2}{I_1}$	0·88	0·86	0·82	0·815	0·79	0·75	0·715	0·67

is not surprising that it is the positions of these lower order resonances which are most affected by the non-uniformity.

For $\rho = 7500$, it is seen that the parabolic law with $\alpha = 0\cdot6$ gives fair quantitative agreement with the experimental data. For this value of ρ, and also for $\rho = 1000$, a variation of the value of α on the position of the first two secondary resonances has been investigated. The results are listed in table 5.2. It is clear that an increase in the absolute value of the density gradient increases the relative spacing between the peaks and that this spacing then tends towards its experimental value.

The next section shows that the use of a better density profile, namely the Tonks–Langmuir one, leads to excellent agreement with experiments for the first two secondary resonances.

Phenomenological picture

Now that the exact nature of the temperature resonances is understood, we describe a phenomenological picture which accounts for the first few secondary resonances. It is based on an order of magnitude calculation by Gould[117]. He uses equation (5-6) which is valid for a uniform plasma but considers

$$k^2(r) = \frac{\omega^2 - \omega_e^2(r)}{\gamma KT/m_e} \quad (5\text{-}15)$$

Perturbed Scalar Pressure Approximation

to be a function of the radius r through the electron density. This is a WBK approximation. By analogy with the standing wave pattern at resonance obtained in the uniform case, he postulates that the integral of the wave number between the radius $r = r_0$ at which it vanishes (the local plasma density $\omega_e(r_0)$ is equal to the frequency ω of the wave) and $r = b$ may have only certain values x_n

$$\int_{r_0}^{b} k \, dr = x_n \tag{5-16}$$

The choice of r_0 is a reasonable one, since for $r < r_0$ we have $k^2 < 0$, i.e. an evanescent wave. With the parabolic profile (5-11), $r_0^2 = b^2(\omega_{e0}^2 - \omega^2)/\alpha\omega_{e0}^2$ and if one takes for x_n the values (5-14) corresponding to the uniform case, one observes a greater separation and a trend toward values in better agreement with experimental results. When $\omega > \omega_{e0}$ the integral is to be taken from $r = 0$ to $r = b$ since k no longer vanishes in the range $0 < r < b$. This phenomenological picture therefore shows that the effect of the equilibrium density gradient is to increase the spacing between successive resonances.

This calculation has been emphasized by Crawford[118,119] in connexion with the concept of series limit introduced by Dattner[96,97]. The latter author had suggested that experimental data similar to that of figure 5.2b indicated that there is a current below which no further secondary resonance exists. The former author proposed a mechanism for the temperature resonances in which he used the order of magnitude calculation described above and limited them to $\omega < \omega_{e0}$, i.e. reflexion must always occur on a plasma core characterized by $\omega < \omega_e$. In this way one predicts only a very limited number of resonances and this seems in agreement with the idea of a series limit. Let us now compare this with the results listed in table 5.1. For values of $(\Omega^2)_{Av}$ smaller than 0·7, $\omega > \omega_e$ anywhere in the non-uniform plasma (since $\alpha = 0·6$) and no resonance can thus be obtained in this region by the mechanism just outlined. All the ten theoretical resonances listed beyond the fifth secondary (the sixth has a $(\Omega^2)_{Av} = 0·69$) necessarily escape the cruder analysis. It will be seen in section 5.2d that many temperature resonances are in fact observed beyond the so-called series limit.

The phenomenological model furnishes nevertheless a good pictorial description of the first few secondary resonances. It must be remembered, however, that the physically fundamental fact is that the relevant boundary conditions are stated at the wall ($u_{r=b} = 0$) and that when a plasma 'core' ($\omega < \omega_e$) exists, it has an effect on the electric field which tends to make the field evanescent in this central region (see also figure 5.6 in the following section).

Finally, comparison between the theoretical results predicted by the approach of the present section and those obtained for the first three secondary peaks by Weissglass[115] using the same profile, shows quite an appreciable difference. The latter author makes the assumption of purely longitudinal waves (curl $\mathbf{E} = 0$ and $\mathbf{B} = 0$); the problem is then restricted in such a way that the main resonance, which is essentially due to the coupling with the electromagnetic field outside, is not included in the treatment. In the quasi-static approximation used here, we take $\mathbf{E} = -\nabla\phi$ but NOT $\mathbf{B} = \mu_0\mathbf{H} = 0$. The meaning of this is assessed in sections 1.3 and 5.2a. The difference between the longitudinal treatment and the present one is essentially that the first leads to a second-order differential equation for $\mathbf{E}$ while the second one leads to a third-order differential equation for $\mathbf{E}$. The influence on the temperature resonances of the restriction to purely longitudinal waves could not be assessed *a priori* and it is interesting to note that this restriction does affect the results.

c. Tonks–Langmuir profile

We have seen in the preceding section that the use of a parabolic profile does not lead to full quantiative agreement between theory and experiments concerning the exact positions of the resonances. The positions of these resonances depend significantly on the density profile used. The solution of equation (5-5) was sought by Parker, Nickel and Gould[79] with a density profile[122] corresponding to the Tonks–Langmuir model[123]. They obtained excellent quantitative agreement for the main resonance and the first two secondary resonances both in the dipolar case and, with suitable excitation (see section 1.10c), also in the quadrupolar case.

The Tonks–Langmuir model of a collisionless plasma in the positive column of a discharge tube has the following characteristics. A Maxwellian electron gas is contained by a radial static electric field. This field is created when the electrons having a high velocity try to reach the wall more rapidly than the massive ions can follow. The ions are assumed to be generated with zero velocity at a rate proportional to the local electron density, and to undergo free fall under the influence of the radial electrostatic field. The boundary condition is simply that the ion current and the electron current cancel each other at the dielectric wall. The density profiles obtained by this model are plotted in figure 5.3 for a mercury plasma; the quantity $b^2/(r_D^2)_{\text{Av}} = m_e(\omega_e^2)_{\text{Av}}b^2/KT$ is used as a parameter. It is related to the parameter ρ used in the previous section by $b^2/(r_D^2)_{\text{Av}} = \rho(\Omega^2)_{\text{Av}}$. Inspection of figure 5.3 shows that these profiles, which take the sheath into account, tend towards the parabolic profile when ρ tends to

Perturbed Scalar Pressure Approximation

infinity. The boundary conditions expressing the continuity of E_θ and D_r at $z = 1$ can be combined under the form of a resonance condition

$$\left[\frac{d\Phi_n(z)}{dz} \bigg/ \Phi_n(z)\right]_{z=1} = -nK_{\text{eff}}; \quad z = r/b \quad (5\text{-}17)$$

where K_{eff} is an effective dielectric constant for the region exterior to the plasma and $\phi = \Phi_n(z)e^{in\theta}$. This K_{eff} takes into account the effects of both glass wall and the multi-pole device in which the plasma column is placed[79]. Equation (5-5) is solved in a fashion similar to that described in section 5.2b. A typical curve giving the value of the logarithmic derivative

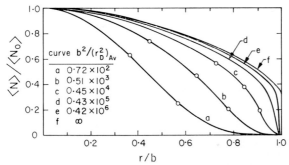

Figure 5.3 Electron density profiles for a mercury plasma derived from the Tonks–Langmuir model (after reference 122)

of the potential as a function of $\omega^2/(\omega_e^2)_{\text{Av}} = (\Omega^{-2})_{\text{Av}}$ is given in figure 5.4. The resonance frequencies are given by the intersection of this curve with $-nK_{\text{eff}}$. It is noteworthy that this curve, and similar curves for other values of the parameter are accurately represented by an expression of the form

$$\frac{\Phi_n'(1)}{\Phi_n(1)} = n\left[1 + \sum_{m=1}^{\infty} \frac{A_{nm}(\omega_e^2)_{\text{Av}}}{\omega_{nm}^2 - \omega^2}\right] \quad (5\text{-}18)$$

The constants A_{nm} and $\omega_{nm}^2/(\omega_e^2)_{\text{Av}}$ are slowly varying functions of the parameter $b^2/(r_D^2)_{\text{Av}}$ and depend also slightly on the multi-pole order n (see table 5.3). The bracketed quantity in equation (5-18) behaves as an apparent equivalent dielectric constant for the mode considered. This quantity satisfies the conditions for a lossless dielectric, and it should be noted that, for the dipole mode ($n = 1$), a sum rule

$$\sum_{m=1}^{\infty} A_{1m} = 1 \quad (5\text{-}19)$$

seems to be obeyed. As stressed by the authors[79], this is not completely unexpected, since it leads to an apparent dielectric constant at high

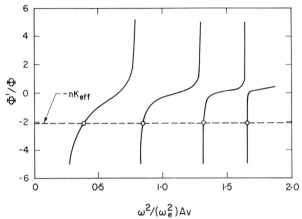

Figure 5.4 Logarithmic derivative Φ'/Φ of the potential as a function of $\omega^2/(\omega_e^2)_{Av}$ for $b^2/(r_D^2)_{Av} = 1600$ (after reference 79)

frequencies ($\omega^2 \gg (\omega_e^2)_{Av}$) which is given by $1 - (\omega_e^2)_{Av}/\omega^2$. Table 5.3 shows that the value of A_{mn} is largest for the lowest or main resonance and decreases monotonically for the higher resonances. This corresponds precisely to the results obtained with the parabolic density profile: as shown by figures 5.1a and 5.1b, the importance of the successive resonances decreases monotonically. Inspection of the values of A_{mn} shows that the importance of the main resonance relative to the secondary ones is greater for relatively high $b^2/(r_D^2)_{Av}$, i.e. when we begin to tend toward the cold plasma limit.

The dipole experimental data for a given discharge tube are shown as circles on figure 5.5 and are compared to the theoretical results shown as solid lines. The electron temperature chosen is that which best fits the data; in this way one does away partially with the problem of determining the best value of γ. The broken line indicates the result obtained for the main resonance by the uniform cold plasma theory (see sections 1.4

Table 5.3 Values of the constants of equation (5-18) for the dipolar mode (from reference (79))

$\dfrac{b^2}{(r_D^2)_{Av}}$	$\dfrac{\omega_{11}^2}{(\omega_e^2)_{Av}}$	A_{11}	$\dfrac{\omega_{12}^2}{(\omega_e^2)_{Av}}$	A_{12}	$\dfrac{\omega_{13}^2}{(\omega^2)_{Av}}$	A_{13}
72	0·6404	0·8521	2·656	0·1297	4·977	0·01859
194	0·3657	0·8412	1·618	0·1370	2·815	0·02177
515	0·2202	0·8507	1·106	0·1208	1·859	0·0254
1580	0·1270	0·8692	0·7983	0·0907	1·302	0·0242
5430	0·0767	0·8876	0·6431	0·0646	1·014	0·0204

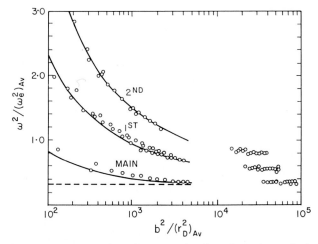

Figure 5.5 Comparison between the first three theoretical dipole resonances and the experimental results shown as circles ($b = 0.5$ cm, $K_{\text{eff}} = 2.1$, $T_e = 33{,}000°K$) (after reference 79)

and 3.1) using a constant ω_e^2 equal to $(\omega_e^2)_{\text{Av}}$. The same computational difficulties that were mentioned in the previous section appear also here. Results could not be obtained for high values of $b^2/(r_D^2)_{\text{Av}}$. (Note that the ρ defined in section 5.2b is linked to $b^2/(r_D^2)_{\text{Av}}$ by $b^2/(r_D^2)_{\text{Av}} = \rho(\Omega^2)_{\text{Av}}$.)

The perturbed electron density resulting from the above calculations is shown as a function of r on figure 5.6 for the first four resonances. The critical point $r = r_0$ corresponding to $\omega_e(r_0) = \omega$ is indicated. It is observed that each higher resonance contains one more half-wavelength of the plasma wave. It requires greater radial distance in which to fit

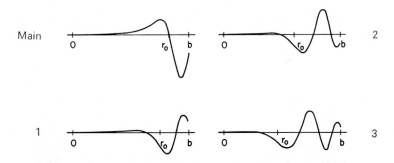

Figure 5.6 Perturbed electron density n for the main and the first three secondary resonances ($b^2/r_D^2{}_{\text{Av}} = 1600$) (from reference 79)

the extra wavelength and, for a given density profile, the incident frequency must be raised in order that more of the plasma can be propagating ($k^2 > 0$); the value of r_0 is thus lowered. These results are, for the first few resonances, in qualitative agreement with the phenomenological model discussed at the end of section 5.2b, but show also that the phenomenon at hand is more involved than is suggested by this simple model. Again, it is clear that the relevant boundary conditions at the wall are the essential factor.

Trial computations have been performed[82] using the consistent description of the perturbed tensorial pressure of section 1.8a. These computations show that the predicted resonance frequencies are practically in as good agreement as those predicted using the scalar pressure approximation but that a 15 per cent higher electron temperature is needed to obtain a good fit to the experimental data. That the best T is not the same in both cases is no surprise since we know that in the scalar pressure approximation it is, in fact, something near to the best fit of γT which is sought (see equation (5-5)). Further work on the comparison between the scalar pressure and the more consistent tensorial pressure approach seems desirable.

Experiments on the quadrupolar mode were also made using the electrode configurations described in section 1.10c. Very good agreement between theoretical results and experimental data is again obtained.

The very satisfactory agreement between theory and experiments obtained by using a density profile resulting from the Tonks–Langmuir model of the positive column of a gas discharge tube shows:

(i) Not only is the over-all behaviour (section 5.2b) of a hot inhomogeneous plasma well described by theory using the truncated moment equations with a scalar perturbed pressure, but precise quantitative results are obtained for the position of the resonances if the proper density profile is chosen.

(ii) The uniform agreement for all values of $b^2/(r_D{}^2)_{Av}$ indicates that the static density profiles calculated from the Tonks–Langmuir model is quite adequate in the experimental conditions used (low pressure mercury discharges).

(iii) The quantities T, or more precisely a value intermediate between T and γT, and $(\omega_e{}^2)_{Av}$ can be satisfactorily deduced from a resonance spectrum and, from these quantities, a radial density profile can be inferred.

(iv) As will be stressed in the general conclusions of section 5.2 the single resonant frequency obtained using uniform cold plasma theory with a $\omega_e{}^2 = (\omega_e{}^2)_{Av}$ is in close agreement with that (the one occurring at lowest density) obtained by the hot non-uniform approach, provided $(r_D{}^2)_{Av} \ll b^2$.

Perturbed Scalar Pressure Approximation

d. Detailed structure of the secondary spectrum

We have seen in section 5.2b that, for $\rho = 7500$, many resonances were theoretically predicted beyond the so-called series limit. These resonances have been experimentally observed (Messiaen and Vandenplas, reference 102). The experimental procedure is that described in section 1.10a. The significant difference is that the reflected signal, after amplification by the receiver, goes through a R-C high pass filter which enhances the rapid variations of this signal as a function of the discharge current $I \propto (\Omega^2)_{\mathrm{Av}}$. In this way the secondary peaks of small amplitude appear preferentially

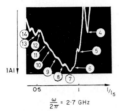

Figure 5.7 Oscillogram giving the 'differentiated' amplitude as a function of I/I_5 (I_5 is the current corresponding to the fifth secondary resonance). The resonances are indicated by arrows

amplified on the oscilloscope screen and can be detected rather easily. Figure 5.7 shows a typical oscillogram obtained by this technique; peaks 4 to 14 are clearly observed. As was seen in figure 5.2b, the first five or six temperature peaks are easily observed by using the conventional techniques. A continuity analysis is performed and the set of secondary peaks is studied as a function of ω. The positions of peaks 5 to 11 as a function of ω show that these high order peaks are not due to spurious or erratic effects.

In a second experiment a small steady magnetic induction $\mathbf{B}_0$ is perpendicular to the axis of the cylinder and parallel to the Poynting vector $\mathbf{S}_{\mathrm{inc}}$ of the incoming plane wave (see figure 4.3). The positions of the secondary peaks are studied as a function of the magnitude of the transverse B_0. A new set of secondary peaks appear and we obtain the intricate pattern of figure 5.8 where the secondary peaks 2–2' to 10–10' are listed as a function of B_0. Note that the peaks of the new set beyond the sixth begin to appear for relatively higher values of B_0 than the first six. Peak 7 has an intermediary behaviour and appears to be on the border line between two regimes. Again, as in section 4.3, the equilibrium density in the cross-section of the cylinder is asymmetrical. This asymmetrical non-uniformity arises from $\mathbf{B}_0$ being perpendicular to the axial discharge current I and the plasma being therefore pushed against the wall in a direction perpendicular to both B_0 and I (see figure 5.9). The effect of asymmetry on the behaviour of the secondary peaks will be investigated theoretically in section 5.2g but we already notice that only a slight asymmetry is needed to generate the new set. When $\mathbf{B}_0$ is perpendicular to

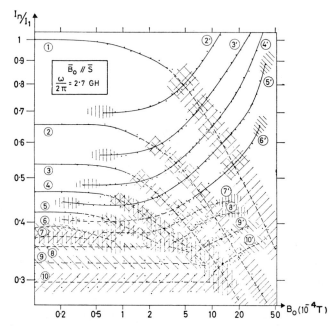

Figure 5.8 I_n/I_1 of the successive secondary resonances as a function of the magnitude of the transverse B_0 (I_1 is the discharge current corresponding to the first secondary resonance when $B_0 = 0$). The 1' curve falls outside the diagram

both the axis and to the Poynting vector $\mathbf{S}_{\text{inc}}$, the direction of asymmetry is no longer along the electric field $\mathbf{E}_{\text{inc}}$ but perpendicular to it and the temperature peaks give no longer rise to a second series but undergo a very marked spreading.

Anyhow, the experimental data of figure 5.8 concerning the influence of $\mathbf{B}_0$ on the secondary peaks beyond the seventh gives conclusive evidence that temperature peaks certainly exist, in the experimental conditions described, up to at least number 11 and that there is no series limit in the sense indicated above. Finally, it should be noted that one resonance beyond the 'series limit' has also been observed in neon by Schmitt[107].

Perturbed Scalar Pressure Approximation

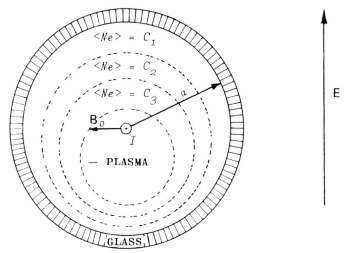

Figure 5.9 Asymmetrical electron density distribution due to the $\mathbf{B}_0 \times \mathbf{I}$ forces

e. Damping effects

Up till now collisional damping effects have been neglected in the theoretical analysis and since the quasi-electrostatic approximation was made, radiation damping (see problem following equation (3-16)) was also excluded and a resonance corresponded to an infinite value of the scattered electric field.

The influence of collision damping has already been investigated for the hot uniform plasma slab-condenser system (section 2.1d). It was noted that the collisions have a much greater influence on the secondary peaks than on the main one. When ν/ω is sufficiently small (ν: collision frequency), the damping of the main, or cold plasma, resonance is preponderantly due to radiation damping. This is the case, for example, with mercury plasmas at pressures of the order of 10^{-3} Torr and with operating frequencies in the GHz range as is seen in section 4.4b. Another phenomenon which can play a role is that of the collisionless damping as is seen in section 2.4*.

Franklin and Bryant have studied in detail[59,65] the damping of the resonances when the cylinder of plasma is placed in a waveguide. The

---

* Landau damping of electron plasma waves propagating along a cylindrical plasma column immersed in a magnetic field has been observed directly by Malmberg, Wharton and W. E. Drummond[129]. The magnitude of the observed damping and its dependence on phase velocity agrees with the theory of Landau damping. Further experimental work on this subject has been done by Derfler and Simoner[214] and Van Hoven[215].

conclusion is that, in mercury, the observed resistive damping of the temperature resonances can not be simply related to the collision frequency ν for momentum transfer. Experiments with gases other than mercury vapour[59,105,106,107] show that the amplitudes of the secondary peaks relative to the amplitude of the main one are significantly smaller than with mercury. No satisfactory and consistent explanation for this has yet been given.

Considerable theoretical attention has been given to the collisionless damping of the temperature spectrum exhibited by the plasma column[115,137,138,139,140]. It has, however, been underlined in section 5.2b that the fact that the experimental resonance peaks become *less* and *less important* (this is not synonymous with *damped*) as electron density decreases, is well accounted for by the fluid theoretical approach. Since this aspect of the situation seems to be often overlooked in the recent studies mentioned, it appears necessary to elaborate this point[141].

The influence of a collision frequency ν on the behaviour of the simple plasma slab-condenser system is analysed on figures 2.4 and 2.5 of section 2.1d. It is clear that a given ν/ω has a much greater influence on the temperature peaks than on the main one. Let us now examine the Q of these resonances. Assuming a Lorentzian resonance profile, it is well known that the response is of the form $K/\sqrt{[\Omega^2 - \Omega_{\text{res}})^2 + (\nu/\omega\Omega^2)^2]}$ where K is the strength of the coupling between the excitation and the response and where $\Omega^2 = \omega_e^2/\omega^2$. With this assumption, the theoretical Q is then given by $\Omega_{\text{res}}/|\Delta\Omega| = \omega_{\text{res}}/|\Delta\omega| = \omega_{e\text{res}}/|\Delta\omega_e|$ and it is immaterial whether ω or ω_e is varied. The real 'measured' Q corresponding to fluid theory is obtained from the theoretical figures 2.4 and 2.5; the difficulty of the measurement lies in the non-zero value of the background signal for the temperature resonances. The profile is no longer Lorentzian and the Q is measured with respect to the background curve as given by $\nu/\omega = 10^{-1}$; the real 'measured' results are compared to ω/ν in table 5.4. For low values of Q and especially for the temperature peaks there exists thus a considerable difference between the $Q = \omega/\nu$ of the Lorentzian profile and the Q measured on the real curve, i.e. the one obtained by fluid theory. The results are particularly striking for $\omega/\nu = 10$.

Table 5.4 Q_0 (main resonance), Q_1 and Q_2 (first and second temperature resonances) as measured on figures 2.4 and 2.5

ω/ν	Q_0	Q_1	Q_2
10^3	10^3	$7{\cdot}46 \times 10^2$	$2{\cdot}1 \times 10^2$
2×10^2	2×10^2	$1{\cdot}9 \times 10^2$	$5{\cdot}9 \times 10$
10^2	10^2	$9{\cdot}5 \times 10$	3×10
10	$8{\cdot}7$	0	0

Perturbed Scalar Pressure Approximation

The experimental resonance curve exhibited by a plasma column is compared to the corresponding theoretical curve obtained by fluid theory on figure 5.2 of section 5.2b. Note that the theoretical resonances correspond to infinity since no radiation or collision damping has been included. The theoretical curve clearly indicates however that the *strength K* of the coupling between the incident field and the resonance decreases with the order of this resonance. Fluid theory therefore predicts that, even if the Q were to remain constant, the resonance peaks become *less* and less *important* as the order of the peak increases. The experimental Q has been

Table 5.5 Experimental values of Q_0 (main peak) and of Q_1, Q_2, ... (successive temperature peaks) for mercury and other gases. T_e is the electron temperature and p is the gas pressure.

Gas	T_e (10·000°K)	p (Torr)	Q_0	Q_1	Q_2	Q_3	Q_4	Q_5
Hg	4	2×10^{-3}	14	27	36	51	57	59
Cs	0·7	10^{-2}	19	23	28	34		
A	8·8	6×10^{-3}	11	11	25	26		
He	7·6	1.5×10^{-1}	8	13	23			
Ne	14	2×10^{-2}	10	10	16			
H_2	7·5	9×10^{-2}	8	15	18			
N_2	7·6	10^{-2}	8	8	5			

measured with respect to the background curve of figure 5.2b which is obtained in Hg and the results are given in table 5.5. These values of Q are trustworthy since great care has been taken to have a linear detection and to obtain a vertical signal on the oscillogram which is rigourously proportional to $|E_{osc}|$. In most other experimental set-ups, it is the Q of the transmitted field, i.e. the combination of incident and scattered field, which is measured. Incidentally, these high Q values do not agree with those given in reference 140 but these have not been obtained by quite the same method.

In table 5.5, the values of Q obtained in other gases are also listed. With the single exception of N_2, the over-all behaviour of Q is the same, i.e. Q increases with the order of the resonance. Further details on the amplitudes of the successive resonances, the collision frequencies, etc. .... are given in reference 105. It is important to remark that it can be shown that radiation damping limits all the peaks to the same amplitude. Indeed if radiation damping is included, equation (5-12) becomes

$$G = \frac{\Delta_G}{\left(\Delta + i\frac{\pi^3}{\lambda_0^2}\Delta_G\right)}$$

provided that the tube radius is much smaller than the vacuum wavelength. At resonance ($\Delta = 0$), $|G|$ is therefore always equal to λ_0^2/π^3 whatever the order of the resonance considered. Since the successive resonances become narrower and narrower, this means that the Q_{rad} due only to radiation damping increases very sharply with the order of the resonance. Now, if one considers the collisions as the only damping phenomenon, the corresponding Q_{coll}, see table 5.4, decreases with the order of the resonance in the simple case of the uniform plasma slab. Combining these two features, it is easily seen that, in the conditions of the experiments described, the Q of the main resonance is essentially determined by the radiation damping. For the first temperature resonance, Q_{rad} is already extremely high and the observed Q is governed by collision and possibly collisionless damping phenomena. Since the real plasma has a strong radial non-uniformity, the simple considerations on the uniform plasma slab give only a rough representation of the collision effects but stress the fact that ω/ν or an equivalent value of it does not suffice to determine the Q of the secondary resonances.

Now, although the experimental values of Q increase with the order of the resonance, these successive resonance peaks diminish in amplitude because the *strength* of the coupling to the incident field decreases. The fact that Q remains high explains, however, why as many as six secondary resonances are observed on figure 5.2b and why a total of eleven to fourteen secondary resonances have been observed by using an appropriate differentiation technique (see section 5.2d). This latter experimental result excludes strong collisionless damping.

f. Noise spectrum

It was seen in section 5.2a that the longitudinal plasma oscillations couple to the transverse electromagnetic waves through the non-uniformity of the plasma or at the interface between a finite plasma and an outside medium. This leads us to expect that the natural longitudinal modes will be thermally excited and will radiate. Measurements on the noise spectrum have been performed and theoretical work done by a number of workers (Agdur, Kerzar and Sellberg[124,125], Lustig[126], Weissglas[127], Gould[128]). These results show that noise radiation exhibits sharp maxima under the same conditions that the scattering maxima exist.

Gould[128] has developed a phenomenological theory which describes adequately the as yet limited experimental observations. Since the radiation is due to the electrons and that the properties of the plasma which involve the ions and neutrals (sheath, collisions, etc. . . .) depend very little on the velocities of the heavy particles, a very good assumption is that the entire

Perturbed Scalar Pressure Approximation

system is in thermodynamic equilibrium at the electron temperature T_e. One may then use Nyquist's theorem to obtain the emission spectrum. To cast the problem in a form to which Nyquist's theorem is applicable, one has to define surface admittances as a function of the magnetic and electric fields. An emissivity is defined which is a function of angular frequency ω and which is equal to unity when it is maximum.

Now, the surface admittance defined in the above sense can be directly related to the theoretical analysis of section 5.2 since it is proportional to formula (5.18) which can be considered as introducing an apparent equivalent dielectric constant for the hot non-uniform plasma. In fact, the surface admittance is given, for the n-th multipolar mode, by

$$Y_n = -\frac{i\omega\epsilon_0 b}{n}\left[1 - \sum_{m=1}^{\infty} \frac{A_{nm}(\omega_e^2)_{\text{Av}}}{\omega^2 - \omega_{nm}^2}\right] \tag{5-20}$$

Equation (5-20) obtained by a collisionless hydrodynamical theory does not include either collisional or Landau damping. By analogy with electric circuit theory and relaxation phenomena, it is reasonable to assume that collisional damping can be introduced by replacing $(\omega^2 - \omega_{nm}^2)$ by $[(\omega - i\nu_{nm})^2 - \omega_{nm}^2]$ in equation (5-20). As seen in section 2.4, Landau damping can also be described in the same way. The dissipation parameters ν_{nm} are, of course, adjustable parameters of this phenomenological approach and must be chosen to fit the experimental results.

The dipolar mode ($n = 1$) is investigated. The values of A_{1m} and ω_{1m} are found in table 5.3. The collision parameter $\nu_{1m}/\sqrt{(\omega_e^2)_{\text{Av}}}$ is assumed to be independent of m and for five values of this parameter, emissivity and reflectivity curves are calculated as a function of $(\Omega^2)_{\text{Av}} = (\omega_e^2)_{\text{Av}}/\omega^2$. For small values of $\nu_{1m}/\sqrt{(\omega_e^2)_{\text{Av}}}$, the reflectivity is essentially independent of the dissipative parameter since the damping is, as seen earlier, primarily due to radiation damping. Emission lines occur at the same values of $(\Omega^2)_{\text{Av}}$ as the reflectivity maxima. These lines are sharp and the main resonance peak $[(\Omega^2)_{\text{Av}} \simeq 2$ since there is no glass wall] has an amplitude which is less than that of the first few secondary peaks. As the dissipative parameter increases, the reflectivity maxima are reduced in amplitude and increased in width as has been underlined earlier. The interesting feature is that the emissivity generally increases as the dissipation parameter is increased until the emissivity at the center of a given peak reaches a maximum, i.e. one, and then decreases again. At the same time the emission peaks broaden and finally merge into a continuum. These results of the phenomenological theory agree well with the experimental data[125]. The observations[125,128] indicate that for those experimental conditions, the dissipative parameters are small and probably in the range $0.001 < \nu_{1m}/\sqrt{(\omega_e^2)_{\text{Av}}} < 0.01$. It is possible, of course, by taking different values of

$v_{1m}/\sqrt{(\omega_e^2)_{Av}}$ for each m, to change the relative amplitudes and width of the emission peaks to a limited extent.

g. Asymmetry effects

When a small steady magnetic induction $\mathbf{B}_0$ is perpendicular to the axis of the cylinder and parallel to the Poynting vector $\mathbf{S}_{\text{inc}}$ of the incoming plane wave (see figure 4.3), it was seen in section 5.2d that a second series of temperature peaks appears and we then have the rather intricate pattern of figure 5.8. It was further noted that this situation is characterized by the fact that, due to the $\mathbf{B}_0 \times \mathbf{I}$ forces, the electron density distribution is asymmetrical along the $\mathbf{E}$ field of the incoming plane wave (figure 5.9).

To investigate the effect of an asymmetrical non-uniformity of the electron density distribution, a simple model consisting of a plasma slab of two unequal regions of uniform but different densities has been studied by Nihoul and Vandenplas[77,130]. The x-axis is taken across the plasma slab. From $x = 0$ to $x = L + a$, the plasma is uniform (region I) and characterized by an equilibrium density $\langle N_1 \rangle$; from $x = L + a$ to $x = 2L$, the plasma is also uniform (region II) but characterized by an equilibrium density $\langle N_2 \rangle$ with $\langle N_2 \rangle$ greater than $\langle N_1 \rangle$. This is a crude but rather fair approximation, as far as the experiments described are concerned, of an asymmetrical parabolic density profile. It must be noted, however, that such a model is very general as it is perhaps the simplest one displaying an asymmetrical inhomogeneity. The plasma slab is excited by a uniform outer electric field E_{ext} of angular frequency ω.

We will use the hydrodynamic equations and it is indicated in section 1.6 that, in the absence of a magnetic field, all the information to first order in the temperature which is contained in the Vlasov equation is also found in the first three hydrodynamic equations. When a steady magnetic field is present, this is no longer true and a full description to first order in the temperature entails the use of the first four moments (0, 1, 2 and 3) of the Vlasov equation[77]. Combining these fluid equations when the steady magnetic induction is perpendicular to the x-axis, one obtains the following equation for the perturbed velocity u_x along x:

$$\frac{d^2 u_x}{dx^2} + k^2 u_x = \frac{\omega^2 - 4\omega_c^2}{3KTe\langle N\rangle/m} J_{\text{Tot}} \qquad (5\text{-}21)$$

with

$$k^2 = \frac{(\omega^2 - 4\omega_c^2)(\omega^2 - \omega_c^2 - \omega_e^2)}{3\omega_e^2 KT/m} \qquad (5\text{-}22)$$

where $\omega_c = eB_0/m_e$ is the electron gyrofrequency. The right-hand side of

Perturbed Scalar Pressure Approximation

(5-21) is proportional to the total perturbed current density $J_{\text{Tot}} = -\langle N \rangle e u_x - i\omega\epsilon_0 E_x = -i\omega\epsilon_0 E_{\text{ext}}$ which is constant because of the one-dimensional geometry. The k^2 given by formula (5-22) is identical to that obtained by Buchsbaum and Hasegawa[131,132] under the same assumptions but by direct use of the Vlasov equation to first order in the temperature. The solution of equation (5-21) is

$$u_x = Q \sin kx + R \cos kx - \frac{i\omega e E_{\text{ext}}}{m(\omega^2 - \omega_c^2 \omega_e^2)} \qquad (5\text{-}23)$$

Note carefully that the k^2 given by equation (5-22) differs, in the absence of a magnetic field ($\omega_c = 0$), from the usual $k^2 = (\omega^2 - \omega_e^2)m/3KT$ given by equation (5-2). The two k^2 differ by a factor ω^2/ω_e^2 and are thus identical to order ρ^{-1}. This is due to the fact that two supplementary moment equations have been used to derive the k^2 of equation (5-22) and the implications of this are carefully analysed in section 1.6.

The hot resonance spectrum of the asymmetrical plasma slab described above is now sought. In each of the regions I and II, the solution (5-23) for the perturbed velocity u_x applies. We, in fact, follow the general lines of section 2.2. We require that u_x be zero at the walls of the slab. We also require that $\langle N \rangle u_x$ be continuous together with its first derivative at $x = L + a^*$. The continuity condition on $\langle N \rangle u_x$ at $x = L + a$ is particularly important from the physical point of view. It must be stressed in this respect that the continuity of $\langle N \rangle u_x$ implies the continuity of E_x (since the total current is necessarily conserved) and thus the continuity of the electric displacement vector since the whole medium is characterized by the permittivity of vacuum ϵ_0 in the present description of a hot plasma. The criticism that, in a cold plasma, each discontinuity in the equilibrium density introduces, through the polarization described by the equivalent permittivities, an additional resonance (see section 1.5a) is thus not applicable to the present hot plasma problem since the necessary precautions are taken at $x = L + a$ so no spurious polarization charges are introduced. Recent work[216] however shows that, even with a hot plasma, a jump discontinuity can introduce a supplementary main resonance.

In the quasi-static approximation, the potential difference between $x = 0$ and $x = 2L$ is equal to

$$V = -\int_0^{L+a} E_{x1} \, dx - \int_{L+a}^{2L} E_{x2} \, dx \qquad (5\text{-}24)$$

---

* The exact boundary conditions are given in the problem on page 209. The last condition is slightly different but reduces to the continuity of the derivative of $\langle N \rangle u_x$ if the asymmetry is small.

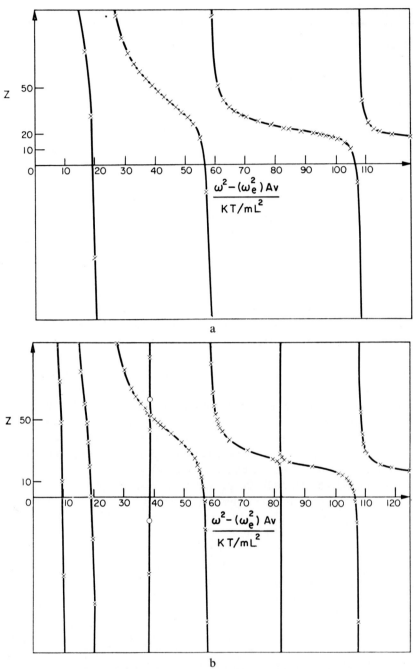

Figure 5.10 (a) Z versus $[\omega^2 - (\omega_e^2)_{\mathrm{Av}}]mL^2/KT$ for the symmetrical plasma ($\nu = 0$) (b) Z versus $[\omega^2 - (\omega_e^2)_{\mathrm{Av}}]mL^2/KT$ for a plasma with an asymmetry of 1 per cent ($\nu = 10$)

Perturbed Scalar Pressure Approximation

and the impedance times the unit surface is given by $Z = V/J_{\text{Tot}}$. Introducing solution (5-23) in equation (5-24), one obtains a complicated expression for Z. Numerical computations of Z as a function of $[\omega^2 - (\omega_e^2)_{\text{Av}}]mL^2/KT$ are carried out for different values of the asymmetry parameter $\nu = \rho(\omega_{e2}^2 - \omega_{e1}^2)/\omega^2$. The results are given in figure 5.10. The figure to the left corresponds to $\nu = 0$; there is no asymmetry and this case corresponds to the plasma slab-condenser system of chapter 2 when the vacuum region is suppressed. The computated values correspond, after allowance for the difference in the k^2 used, to the values obtained analytically and given by formula (2-30). The figure to the right corresponds to a slight asymmetry of 1 per cent. It is clear that this asymmetry introduces a *second* set of temperature resonances. We now investigate the fundamental reason for this.

The impedance Z can be written (see reference 77):

$$Z = F(\omega)\left[1 - \frac{\psi(\omega)}{\Phi(\omega)}\right] \tag{5-25}$$

(1) *Case $\nu = 0$*

The zeros of Φ are given by:

$$kL = (2N + 1)\pi/2; \quad N = 0, 1, 2, \ldots \tag{5-26}$$

$$kL = N\pi; \quad N = 0, 1, 2, \ldots \tag{5-27}$$

It can be readily verified that $kL = N\pi$ is also a zero of ψ and that

$$\lim_{kL \to N\pi} \frac{\psi}{\Phi} = 0 \quad \text{when} \quad N \neq 0$$

Hence Z remains finite and different from zero when $kL = N\pi$ and these values of k do not correspond either to resonances or anti-resonances. The only anti-resonances are thus given by $kL = (2N + 1)\pi/2$ as found in section 2.1 ($2L = a$).

(2) *Case $\nu \neq 0$*

When the density distribution is asymmetrical, the characteristic phenomenon is that ψ and Φ no longer have a common series of zeros (i.e. the zeros that were common for $k = N\pi$ when $\nu = 0$). This gives rise to a new set of anti-resonances ($\Phi = 0$) and thus in their neighbourhood to a new series of resonances. The comparison of curve (b) with curve (a) of figure 5.10 indicates clearly that there exist twice as many resonances in the asymmetrical case ($\nu = 10$) as in the symmetrical case ($\nu = 0$).

The physical reason for this is clear. Since the boundary conditions are symmetric, E is symmetric if the density profile is symmetric. An asymmetric density profile is therefore necessary in order that E also have an anti-symmetric part and it is this anti-symmetric contribution which is responsible for the fact that the zeros of ψ and Φ no longer coincide for $kL = N\pi$ and that a second series of resonances thus appears. It can be thought that this plasma slab behaviour should explain the second series of temperature resonances experimentally observed in the cylindrical case. In the latter case, if the excitation is purely dipolar ($n = 1$) and the plasma radially symmetric, the scattered field only exhibits dipolar resonances. If the plasma is, however, rendered radially asymmetric by the transverse magnetic field, this asymmetry couples the $n = 0$ (radially symmetric) and the $n = 2$ (quadrupolar) modes of the scattered field to the dipolar excitation (see section 4.3 in the cold plasma case) and it is known, for instance, that the quadrupolar temperature resonances do not coincide with the dipolar ones[79,82]. In fact, it is observed that the second series of temperature peaks has practically a $n = 0$ azimuthal dependence.

Finally, the results of the present paragraph can be extended by means of the Liouville asymptotic method valid for large values of ρ to any asymmetrically non-uniform density profile of the plasma[133].

h. Axial propagation effects

Although, as stated in section 1.1, the monograph does not deal with axial propagation effects, it was felt interesting to include a section which clearly shows the connexion between the phenomena studied in this chapter and axially propagating modes. These modes have been studied in a hot non-uniform plasma by O'Brien, Gould and Parker[134,135].

A first set of experiments is carried out to study the rotational symmetric modes ($n = 0$). The wave is excited by the fringing field of a coaxial re-entrant cavity (this cavity posseses rotational symmetry) such as that often used with klystrons and another cavity of this type is fitted on the plasma column to act as receiver. The frequency is kept constant and, as usual, the density of the plasma is varied by modulating the discharge current. The amplitude of the received signal shows many distinct transmission bands between certain values of $(\omega_e^2)_{Av}/\omega^2$. A large variation of the axial wavelength λ_L is observed within each transmission band. At the end of the band corresponding to the highest values of $(\omega_e^2)_{Av}/\omega^2$, the wavelength becomes large and generally of the same order of magnitude as the vacuum wavelength λ_0 whereas at the other end of the band, the wavelengths have a minimum of the order of 0·5 cm. The termination characterized by high values of λ_L corresponds clearly to a cut-off condition ($\lambda_L = \infty$). There exists a predominant transmission domain for high values of $(\omega_e^2)_{Av}/\omega^2$ and this transmission is also predicted by cold plasma theory.

Perturbed Scalar Pressure Approximation

A set of experiments conducted to investigate the dipole mode ($n = 1$) leads generally to results similar to those obtained for the $n = 0$ mode with the exception of the structure of the predominant transmission band occurring at the highest values of $(\omega_e^2)_{Av}/\omega^2$. This latter transmission band is also, as expected, predicted by cold plasma theory[136], i.e. by solution of $\nabla \epsilon_0 (1 - \omega_e^2/\omega^2) \nabla \phi = 0$, and it is seen that the inclusion of a finite electron temperature T_e modifies the theoretical behaviour only slightly in the region of large wavelengths. The temperature resonances studied in the previous sections correspond naturally to the cutoffs at the high $(\omega_e^2)_{Av}/\omega^2$ end of the less important transmission bands.

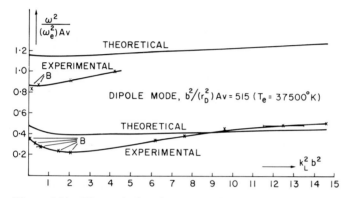

Figure 5.11 Theoretical and experimental curves giving $\omega^2/(\omega_e^2)_{Av}$ as a function of $k_L^2 b^2$ for the dipole mode. After reference 134

The problem is investigated theoretically for the $n = 1$ mode by using the same hot non-uniform plasma column as in section 5.2c, but one must now also consider a Z-dependence; this extension is straight-forward. The theoretical results are compared to the experimental ones on figure 5.11. The experimental results have been corrected for the drift velocity from cathode to anode of the electrons by comparing the curves obtained by propagation in anode–cathode and cathode–anode directions. One notes that there is a region of backward propagation which is noted B. Such a region is characterized by the fact that phase velocity and group velocity have opposite signs.

Finally, with $n = 1$ excitation and with a small steady magnetic field $\mathbf{B}_0$ perpendicular to $\mathbf{E}$ and to the axis of the column, a second series of temperature transmission bands is observed; this second set of transmission bands coincide with the temperature transmission bands of the $n = 0$ mode. This should be immediately compared to the asymmetrical results (second series of secondary resonances) studied in section 5.2g.

i. Conclusions

Sections 5.2b and 5.2c show clearly that the characteristics of the main resonance obtained from *uniform cold* plasma theory using an ω_e^2 equal to the $(\omega_e^2)_{\mathrm{Av}}$ are in good agreement with the results obtained with the much more elaborate and realistic theoretical model of a hot non-uniform plasma provided $(r_\mathrm{D}/b)^2 \ll 1$. This explains *a posteriori* why the main features of the behaviour of the experimental plasmas of chapters 2, 3 and 4 are so adequately described by finite models of uniform cold plasmas.

When the plasma is in contact with a metallic surface (see chapter 6 and section 2.3), it is again very adequately described by a uniform cold plasma but one *cannot* neglect the sheath region with a thickness of the order of the Debye length r_D. This sheath region is characterized by a low electron density and can thus be approximated by a vacuum layer of permittivity ϵ_0. The role of this vacuum layer is essential in predicting the correct resonant behaviour of the plasma since, with its positive permittivity, it can constitute a resonant circuit with the plasma which displays a negative permittivity ϵ_p when $\omega < \omega_e$.

All the details of the secondary spectrum are well described by the hydrodynamic equations of a *hot non-uniform* plasma. The introduction of radiation damping and of a phenomenological collision frequency in this theoretical description suffice to explain the over-all damping of the successive temperature resonances as well as the fact that so many secondary resonances can be observed when $(r_\mathrm{D}/b)^2$ is very small and if sufficient care is taken. Asymmetry effects are shown to introduce a second set of temperature resonances.

5.3 Influence of steady axial magnetic fields

a. Low magnetic fields ($\omega \gg \omega_c$)

Experiments with a steady magnetic induction $\mathbf{B}_0$ parallel to the axis of the column indicate that when the field is small ($\beta = \omega_c/\omega < 0.2$; ω_c: cyclotron frequency), the temperature peaks broaden somewhat and then disappear when the magnetic field becomes too high[44] but that the positions of the resonances are but slightly affected. It can be shown theoretically[82,142], however, that the temperature peaks should undergo a splitting similar to that observed for the main peak and described in section 4.2b. It is not clear whether the magnitude of this effect of the magnetic field should cause an observable splitting of the temperature resonances; anyhow such a splitting has not yet been conclusively observed. It is important to stress in this connexion that the second set of temperature peaks which arises when B_0 is perpendicular to the axis is already observable when $\beta = 0{\cdot}001$ (see section 5.2d) while the peaks do not split in an axial magnetic field of the order of $\beta = 0{\cdot}1$. This order of magnitude difference between the parallel and perpendicular effects enables one to align the axis of the plasma column parallel to the field by a method of zero effect consisting in the non-observance of the influence of moderate values of B_0 on the temperature peaks.

b. High magnetic fields ($\omega = n\omega_c$; $n = 2, 3, 4, \ldots$)

The microwave absorption in a plasma column near the harmonics of the electron cyclotron frequencies have been studied by Buchsbaum and Hasegawa[131,132]. They observe (see figure 5.12) a structure consisting of many resonance peaks at magnetic fields somewhat higher than the cyclotron harmonic fields corresponding to $\omega_c = \omega/n$ (n: integer). This structure is superimposed on a background absorption which is only a weak function of magnetic field. The details of the structure are more easily discernible near the second harmonic but are also observable with diminishing amplitudes near the third, fourth and fifth harmonics. They theoretically show that the observed resonances are due to the excitation of longitudinal plasma oscillations perpendicular to the magnetic field near each cyclotron harmonic. Such a spectrum was first observed in emission[143] and also further studied in absorption[144,145,146,149]. This detailed spectrum in not the one of Landauer's experiment[147] in which he observed radiation peaks at the cyclotron harmonics up to the 45th but no peaks in the neighbourhood of a given harmonic. These main peaks at the cyclotron harmonics have been satisfactorily explained by Canobbio and Crocci[217].

Buchsbaum and Hasegawa have employed the linearized Boltzmann equation and have integrated it formally[148] using the spatial derivative $(w/\omega_c)\nabla$ as an operator where w is the root mean square velocity. This operator is proportional to the electron temperature T. If the terms retained in the expansion resulting from the integration of the Boltzmann equation are only those which are linear in T and if the quasi-static approximation $\mathbf{E} = -\nabla\phi$ is made, one obtains the following equation for ϕ in a uniform plasma:

$$(\nabla^2 + k^2)\nabla^2\phi = 0 \tag{5-28}$$

with

$$k^2 = \frac{(\omega^2 - 4\omega_c^2)(\omega^2 - \omega_c^2 - \omega_e^2)}{3\omega_e^2 KT/m} \tag{5-29}$$

Note carefully that the k^2 obtained by this method is identical to the one given by equation (5-22) in section 5.2g. The latter expression of k^2 had been obtained by the moments' method using all the terms which are of first order in the temperature. This shows the equivalence between the two methods in obtaining the relevant plasma equations. Such a k^2 is adequate in describing the behaviour of a uniform plasma up to $\omega_c = 2\omega_c$. In order to describe it around the harmonic $\omega = 3\omega_c$, we must retain all the information derived from the Vlasov equation up to T^2. For higher harmonics, we must retain higher powers of T and a systematic scheme to carry this out has been devised by Azevedo[150].

With the further assumption that the oscillations are purely longitudinal (it is underlined in section 3.1a that this is far more stringent than the quasi-static approximation), one obtains the following eigenvalues in a plasma slab of width $2L$

$$k_N^2 = \frac{(\omega^2 - 4\omega_c^2)(\omega^2 - \omega_e^2 - \omega_c^2)}{3KT\omega_e^2/m} = \left(\frac{N\pi}{2L}\right)^2 \quad (5\text{-}30)$$

with $N = 1, 2, 3, \ldots$. We remarked in section 5.2g (case $\nu = 0$), however, that all values of N do not correspond either to a resonance or an antiresonance; only uneven values of the form $N = 2M + 1$, which correspond to a symmetric electric field in the plasma slab, are characteristic of an antiresonance while the even values $N = 2M$, which would correspond to an antisymmetric electric field in the slab, cannot give rise to a resonance in a symmetric plasma. A formula of the type (5-30) is not quite sufficient to account for the experimental results around $\omega = 2\omega_c$ because it predicts too small a spacing between the absorption peaks. This is exactly the same situation as for the ordinary temperature resonances when $\omega_c = 0$ (section 5.2). If account is taken of the variation of the equilibrium density along the x-axis of the slab, i.e. $\omega_e^2 = \omega_{e0}^2 \, g(x)$ where ω_{e0} is the plasma frequency at the centre of the slab, but if the static field $\langle E \rangle$ balancing the density gradient and given by equation (1-105) is neglected, the following equation is obtained in the one-dimensional case

$$\frac{d}{dx}\left[\frac{d^2}{dx^2} + \frac{(\omega^2 - 4\omega_c^2)(\omega^2 - \omega_c^2)}{3KT\omega_{e0}^2/m}\left(\frac{1}{g(x)} - \frac{\omega_{e0}^2}{\omega^2 - \omega_c^2}\right)\right]g(x)\frac{d\phi}{dx} = 0$$

$$(5\text{-}31)$$

If computations are done with equation (5-31), far better agreement is obtained between the theoretical results of the slab and the actual experimental results obtained in cylindrical geometry.

The same problem has also been solved in cylindrical geometry and the results are shown on figure 5.13 where the temperature T has been used to fit the theoretical results (solid curves) to the experimental results.

The dispersion relation (5-29) indicates that $k^2 > 0$ when $\omega < 2\omega_c$

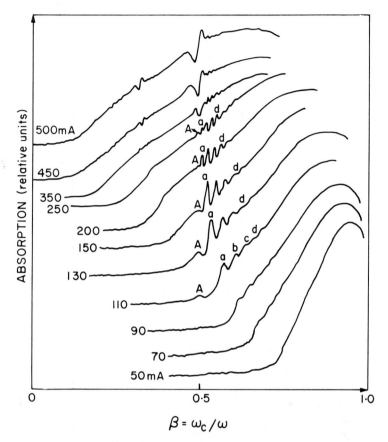

Figure 5.12 Microwave absorption in a helium plasma column as a function of steady axial magnetic field ($\beta = \omega_c/\omega$) with discharge current I_{dis} as a parameter. The curves for different I's are displaced for display purposes (after reference 132)

and $\omega^2 < \omega_e^2 + \omega_c^2$. This condition can be satisfied in a non-uniform plasma column in the central region, i.e. where the density is the highest. The dispersion relation also indicates that $k^2 > 0$ when $\omega > 2\omega_e$ and $\omega^2 > \omega_e^2 + \omega_c^2$ and this condition corresponds to modes located in the sheath region near the column wall. The results of figure 5.12 show that the modes which are mostly observed are those corresponding to the 'standing' waves in the core of the plasma column. The 'sheath' modes are observed at low densities, see section 5.3c and figure 5.14.

The Bauchsbaum–Hasegawa resonances correspond in fact to some of the modes discussed by Bernstein[151] for electrostatic—or longitudinal—oscillations perpendicular to a uniform steady magnetic field in an infinite

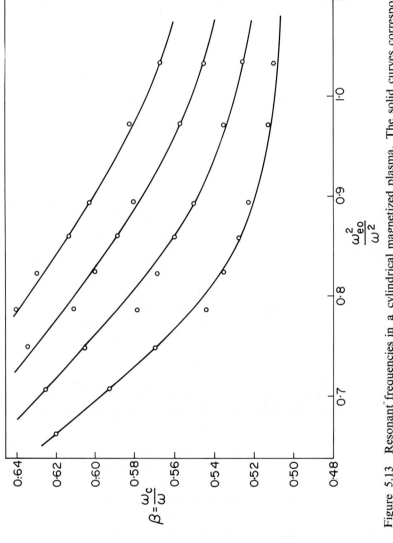

Figure 5.13 Resonant frequencies in a cylindrical magnetized plasma. The solid curves correspond to theory; the points are experimental and correspond to the absorption peaks of figure 5.12 (after reference 132)

Influence of Steady Axial Magnetic Fields

homogeneous plasma. A more complete theoretical discussion of these resonances includes the electrostatic field balancing the plasma non-uniformity as well as the anisotropy of the unperturbed electron velocity distribution (Pearson[152], Weenink, Rem[153]). In the latter of these two works, the equations of the inhomogeneous magnetized plasma have been derived by two methods: one consisting of the integration of the Vlasov equation along the unperturbed orbits of the particles and the other starting from the moments' equations.

A last point is the coupling of the longitudinal wave to the radiation field, i.e. the coupling between the longitudinal component of the electric field and its transverse component. This problem has been examined by Stix[155], Horton[156] and particularly in relation to the experiments described above by Bernstein and Weenink[157] and Kuehl[158]. This must be done for a bounded plasma in the usual fashion: the field inside the plasma is matched to the field outside which has the proper behaviour at infinity compatible with Maxwell's equations.

c. Continuity between low and high magnetic fields ($\omega_c \ll \omega$ to $\omega_c = \omega/2$)

The behaviour of temperature plasma resonances in intermediate axial magnetic fields has been investigated by Schmitt, Meltz and Freyheit[145] in an afterglow neon plasma. The plasma density decays in the afterglow as a function of time and it is assumed that the electrons are substantially in thermal equilibrium with the neutral gas ($T_e = 300°K$). The resonance peaks are listed in figure 5.14 as a function of time t in the afterglow (i.e. decreasing densities) and normalized magnetic field. $\beta^{-1} = \omega/\omega_c$. Such a pattern is observed in a non-uniform plasma when $\omega_e^2(r) + \omega_c^2 < \omega^2$ ($0 < r < b$) and corresponds thus to the plasma region near the wall, see discussion on page 193.

For weak magnetic fields, the behaviour is that described in section 5.3a. When the magnetic field increases, the temperature resonances remain observable and this is ascribable to the small collision frequencies resulting from the typically low electron temperature of the afterglow. This leads to a rather characteristic pattern in which, when the magnetic field is increased beyond a threshold value, each temperature resonance is shifted towards lower densities and then attenuated until it reappears at a higher density*. Note that the resonances at higher magnetic fields occur for $\omega/\omega_c > 2$ since $\omega^2 > \omega_e^2(r) + \omega_c^2$ [see equation (5-29)].

The phenomena observed can be explained semi-quantitatively on the basis of the linearized Vlasov equation by using a WKB approximation[145].

---

* This behaviour is consistent with the dispersion curves of the Barnstein modes as computed by Dreicer[218].

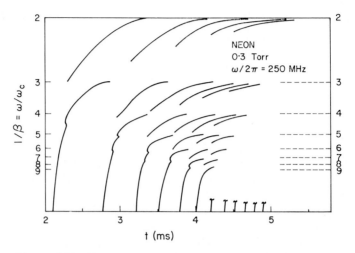

Figure 5.14 Temperature resonances in a cylindrical afterglow plasma as a function of time t in the afterglow (i.e. decreasing plasma density) and normalized axial magnetic field $\beta^{-1} = \omega/\omega_c$ (after reference 145). Note that $\omega^2 > \omega_e^2(r) + \omega_c^2$ with $0 < r < b$

5.4 Non-linear effects

Only linear effects have been considered in the previous chapters and in the previous sections of the present chapter. Although it is not our aim to go very deeply into the study of non-linear phenomena, we will review some work on non-linear effects which is closely connected to the temperature resonances studied in sections 5.1 and 5.2.

a. Non-linear temperature resonances at low power

By beaming a wave on a plasma cylinder in the same conditions (see section 1.10b) that lead to the observation of the ordinary temperature spectrum, Stern[160] has observed the generation of the second harmonic of the excitation signal at the densities for which linear resonance peaks exist. He has also observed the generation of frequencies which are the sums of the fundamental frequencies of two simultaneously excited oscillations.

The measurements are performed at fixed values of the power and of the frequency of the incident signal while the plasma density is varied in the usual fashion. Figure 5.15a shows an oscillographic record of the second-harmonic signal generated at a frequency $2\omega/2\pi = 8\text{GHz}$ when an incident signal at frequency $\omega/2\pi = 4\text{GHz}$ is sent on the plasma column.

Non-linear Effects

The broken line shows the power reflected at the incident frequency ω. When the power of the incident signal is increased, the level of the 8 GHz signal increases sharply within the resonant density ranges but no second-harmonic signals is detected outside the resonant ranges. It is quite clear that the second harmonic signal 2ω is only generated at those electron

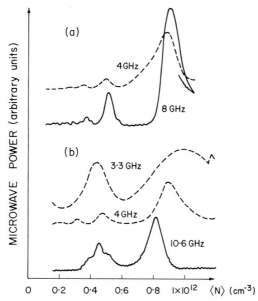

Figure 5.15 (a) Second harmonic signal (solid line) as a function of plasma density $\langle N \rangle$; the broken line gives a simultaneous record of the power of the reflected signal at the excitation frequency. (b) Side-frequency generation of a signal at 10·6 GHz as a function of plasma density $\langle N \rangle$. The broken lines give the simultaneous records of the reflected power at the two excitation frequencies 3·3 and 4 GHz (after reference 160)

densities for which there exist resonances of the incident signal at frequency ω. The electron density represented in abscissa is determined by means of a microwave cavity coaxial with the plasma column (section 1.11).

It is checked that the power of the second harmonic-peak is proportional to the square of the incident power.

Figure 5.15b indicates what happends when two signals of frequencies ω_1 and ω_2 are incident on the plasma: signals at the sum frequency $\omega_1 + \omega_2$ and side frequency $2\omega_1 \mp \omega_2$ are generated. Figure 5.15b gives in fact a typical side-frequency observation. The solid line shows a 10·6 GHz signal generated when two signals at 3·3 and 4 GHz are incident on the plasma and the two broken lines show the power reflected at each incident frequency. We observe that the $10 \cdot 6 = 4 + 2 \times 3 \cdot 3$ GHz signal is generated whenever resonances at the two incident frequencies 3.3 and 4 GHz nearly overlap. In addition to the signal at 10·6 GHz, signals at the sum frequency 7·3 GHz and at the other side frequencies are also observed.

Such observations are found to be consistent with solutions of the nonlinear Vlasov equation; the agreement between measurement and theoretical estimates indicates that the process described represent the inherent non-linear properties of the thermal 'longitudinal' oscillations.

b. Non-linear resonances of a cold plasma at low power

One also expects that nonlinear signals be generated that are linked to the main or cold plasma resonance which always occurs at $\omega\theta$ ($0 < \theta < 1$) in a bounded plasma (see chapters 2 and 3). Such non-linear signals ought to be adequately described by the simpler cold plasma theory instead of the more elaborate theory of the hot non-uniform plasma which is needed to explain the temperature resonances.

Theory and experiments of weak non-linear interactions at resonance in a cylindrical plasma described by cold plasma theory shows that these interactions are also understood (see Messiaen and Vandenplas, reference 161 and 219).

In the cold plasma approximation, the plasma is described by the following equations for the perturbed quantities $\mathbf{u}$, $\mathbf{E}$, $\mathbf{B}$

$$\frac{\partial \mathbf{u}}{\partial t} + (\mathbf{u} \cdot \nabla)\mathbf{u} = -\frac{e}{m}(\mathbf{E} + \mathbf{u} \times \mathbf{B})$$

$$\frac{\partial N}{\partial t} + \nabla(N\mathbf{u}) = 0; \quad \text{curl } \mathbf{H} = \epsilon_0 \frac{\partial \mathbf{E}}{\partial t} - eN\mathbf{u} \quad (5\text{-}32)$$

$$\text{div } \mathbf{E} = \frac{e(\langle N \rangle - N)}{\epsilon_0}; \quad \text{curl } \mathbf{E} = -\frac{\partial \mathbf{B}}{\partial t}$$

where N is the total electron density and $\langle N \rangle$ the equilibrium density (i.e., in the absence of a perturbation signal). Elimination between the different equations of system (5-32) leads to ($\omega_{e0}^2 = e^2 \langle N \rangle / m\epsilon_0$):

$$\text{curl curl } \mathbf{E} + \epsilon_0 \mu_0 \left(\frac{\partial^2}{\partial t^2} + \omega_{e0}^2 \right) \mathbf{E}$$
$$= -\mu_0 \epsilon_0 \left\{ \omega_{e0}^2 \left[(\mathbf{u} \times \mathbf{B}) + \frac{m}{e}(\mathbf{u}\nabla)\mathbf{u} \right] + \frac{\partial (\mathbf{u} \, \text{div } \mathbf{E})}{\partial t} \right\} \quad (5\text{-}33)$$

$$\frac{\partial}{\partial t} \left\{ \text{curl curl } \mathbf{H} + \epsilon_0 \mu_0 \left(\frac{\partial^2}{\partial t^2} + \omega_{e0}^2 \right) \mathbf{H} \right\}$$
$$= \epsilon_0 \left\{ \frac{\partial}{\partial t} \text{curl } [(\text{div } \mathbf{E})\mathbf{u}] + \omega_{e0}^2 \left[\text{curl } (\mathbf{u} \times \mathbf{B}) + \frac{m}{e} \text{curl } [(\mathbf{u} \cdot \nabla)\mathbf{u}] \right] \right\}$$
$$(5\text{-}34)$$

Non-linear Effects 169

When $\partial/\partial Z \equiv 0$, equations (5-33) and (5-34) are solved to first order, i.e. neglecting the non-linear righthand side, and one first order solution in $e^{i(n\theta - \omega_1 t)}$ and another one in $e^{i(s\theta - \omega_2 t)}$ are then introduced in the righthand side which acts as source term for the non-linear solutions E_θ and H_Z. These fields are matched at $r = b$ to the fields existing in the glass wall of

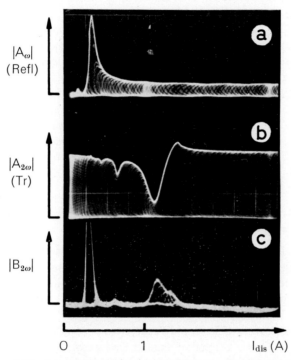

Figure 5.16 (a) Amplitude $|A_\omega|$ of the 'linear' reflected signal at ω as a function of $I_{\text{dis}} \propto \langle N \rangle$. (b) Amplitude $|A_{2\omega}|$ of the 'linear' transmitted signal at 2ω as a function of $I_{\text{dis}} \propto \langle N \rangle$. (c) Amplitude $|B_{2\omega}|$ of the non-linear signal generated at 2ω as a function of $I_{\text{dis}} \propto \langle N \rangle$

the plasma column and at $r = c$ to the fields in the vacuum outside. One obtains in this way the amplitude $C_{|n+s|}$ of the scattered non-linear field in $e^{i(\omega_1 + \omega_2)t}$ which exhibits three resonances.

The experiments are characterized by plane wave excitation and since $c \ll \lambda_0$, the $n = s = 0$ and $n = s = \pm 1$ modes are preponderant in the excitation. C is therefore computed as a function of plasma density for $\omega_1 = \omega_2 = \omega$ and suitable combinations of $n, s = 0, +1, -1$ leading to C_1 and C_2. These theoretical C_1 and C_2 display exactly the resonances which are observed experimentally and which are described.

A plasma column is placed between an emitting waveguide and a receiving waveguide having a cut-off frequency greater than ω. The 'linear' reflected field $|A_\omega|$ at ω as measured through the emitting waveguide is shown on figure 5.16a. When a wave at 2ω is sent on the plasma, the 'linear' transmitted field $|A_{2\omega}|$ at 2ω as observed through the receiving waveguide is shown on figure 5.16b. The smaller absorption peaks to the left are temperature resonances.

The field $|B_{2\omega}|$ at 2ω generated by non-linear interaction in the plasma is shown on figure 5.16c. The sharp peak to the left on this figure indicates non-linear scattering at the main resonance density corresponding to ω (figure 5.16a). The broad peak to the right of figure 5.16c indicates non-linear scattering at densities corresponding to a combination of the main dipolar resonance at 2ω and the main quadrupolar resonance at 2ω (see figure (5.16b)). The features of this field generated at 2ω by non-linearities are predicted theoretically. The theoretical amplitudes are of the order of 110 dB below $|A_\omega|$ and the observed amplitudes are of the order of 95 ± 10 dB. It is checked that $|B_{2\omega}| \propto |A_\omega|^2$.

c. Microwave scattering from density fluctuations resulting from temperature resonances

Stern and Tzoar[163] have observed microwave scattering from the enhanced electron density fluctuations resulting from temperature resonances excited in a plasma column. The problem of incoherent scattering from fluctuations in a plasma has been studied theoretically by a number of authors[164-170].

In the experiments of Stern and Tzoar, two microwave signals are beamed at the plasma. One is a signal of angular frequency ω_{inc}—the scattering of which is being studied—and the plasma density is such that the plasma frequency on the axis of the column is much smaller than ω_{inc}. The other signal—the resonance signal—is characterized by ω_{res} and can excite a temperature resonance in the plasma (section 5.2). The scattering of the microwave signal at ω_{inc} from the plasma leads to the appearance of a scattered signal at $\omega_{\text{sc}} = \omega_{\text{in}} - \omega_{\text{res}}$ whenever the density of the plasma is such that it corresponds to a temperature resonance at ω_{res}.

Figure 5.17 shows a typical result of such an experiment ($\omega_{\text{inc}}/2\pi = 10.5$ GHz; $\omega_{\text{res}}/2\pi = 2.6$ GHz; $\omega_{\text{sc}}/2\pi = 7.9$ GHz). Trace (a) gives the reflected power at 2·6 GHz as a function of ω_e^2 and it exhibits the usual temperature resonance spectrum. Trace (b) displays the corresponding scattered power at 7·9 GHz and the scattered signal is only observable in those ranges of ω_e^2 at which a plasma resonance indicated by (a) is present. This effect cannot be due to the frequency mixing examined in section 5.4a

Non-linear Effects

since $\omega_{inc} \to \omega_e$ and that the amplitudes of the temperature resonances at ω_{inc} in the ω_e^2 domain considered are thus extremely small. In fact, the experimental results are found to be quite consistent with the theory of scattering from fluctuations if the perturbed plasma density $n(\mathbf{r}, t)$ is taken to be that given by the theoretical approach of section 5.2 and which is considerably enhanced at resonance.

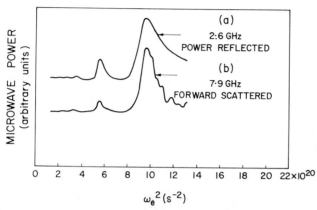

Figure 5.17 Traces of (a) reflected power at $\omega_{res}/2\pi = 2 \cdot 6$ GHz, (b) forward-scattered power at $\omega_{sc} = \omega_{inc} - \omega_{res}$ as a function of ω_e^2 when $\omega_{inc}/2\pi = 10 \cdot 5$ GHz. Zero level of traces is given by the reading at $\omega_e^2 \to 0$ (after reference 163)

d. Strong non-linear effects and resonantly sustained plasmas

If, in the situation experimentally studied in section 1.10b (plasma column inserted across a waveguide) and theoretically analysed in sections 5.2a, 5.2b and 5.2c, the incoming power is increased above 0·5 W, characteristic strong non-linear, effects appear: deformation of the resonance peaks and hysteresis phenomena[205,206,207,208]. Since they are not amenable to a second order treatment, these last effects differ of course fundamentally from the weak non-linear ones studied in sections 5.4a and 5.4b.

Using an independent measurement of electron plasma density, one can show that, as the power is further increased, there is a preferential absorption of energy at resonance which causes high frequency ionization and a corresponding alteration of the zero order density[206]. This results in a tendency of the plasma to remain in a resonance state even when the discharge current is varied. When the incident power is sufficiently high, the high frequency energy absorbed at resonance *sustains* the plasma in a state corresponding to a resonance at ω even in the absence of an external discharge current. This phenomenon should be related to the fact that a

HF plamoid is a plasma at resonance (see section 2.3). If the plasma is in such a resonant state and if the incoming HF power is decreased below a certain minimum, a jump to another resonance occurs characterized by a lower density.

When the plasma is fastened on a resonance, the plasma density does not drop suddenly to zero outside the waveguide, there exsist plasma 'tongues' with length and luminosity which increase with incoming power. This last feature is consistent with the assumption that the plasma is accelerated outside the waveguide region by a mean force which is proportional to $-\nabla|\mathbf{E}|^2$ [209,210,211] since such a gradient of electric field exists along the axis of the column.

The influence of a steady axial magnetic field $\mathbf{B}_0$ on the resonantly sustained behaviour of the plasma has also been studied[212]. When the incoming power is high enough (of the order of 35W or more), the observed results are independent of this power. A striking pattern for the density versus ω_c/ω ($\omega_c = eB_0/m_e$) is observed. This pattern shows the role of the cyclotron harmonics and also the extinction of the plasma at $\omega = \omega_c$. A clear semi-quantitative theory can be established since in first and quite good approximation the possible modes in the magnetized plasma column are the electromagnetic cold modes together with the modes of the infinite hot plasma[151] having the relevant eigenvalues of k^2 imposed by the boundary conditions. The plasma is expected, at high HF power, to be sustained in that mode which is characterized by the highest density compatible with the coupling of the resonance to the external field.

Since this topic falls somewhat outside the scope of this book, no further detail concerning these strong non-linear phenomena will be given and the interested reader is referred to the original literature.

CHAPTER 6

Metallic and Dielectric Resonance Probes. Plasma-Dielectric Coated Antenna. General Considerations

This chapter is devoted to applications of the resonance properties of plasma systems studied previously. We first examine the metallic resonance probe which is essentially a suitable biased Langmuir probe with a high-frequency signal superimposed on it. The properties, at resonance, are found to be strongly dependent on the thickness of the sheath surrounding the probe. The dielectric resonance probe is then introduced; it is constituted by a dielectric rod containing two wires to which a high-frequency signal is applied. The resonant properties of the dielectric probe and the plasma density measured are, to a large extent, insensitive to the thickness of the sheath. The last system to be considered is an antenna coated with a dielectric and a plasma layer. The plasma sheath (or dielectric layer) existing between this antenna and the surrounding plasma layer plays an important role. It is predicted that, for a given operating frequency, there exist certain plasma densities for which the system is resonant and this very strongly enhanced radiation is fully confirmed by experiments.

Finally, general considerations are given on the difference between the resonant, and anti-resonant, behaviour of plasma slab-condenser systems, metallic resonance probes and plasma-coated antennas, i.e. systems having at least one closed metallic boundary, on one hand, and that of plasma cylinders and dielectric resonance probes, i.e. systems having no metallic boundary, on the other hand. The mathematical reasons for the similarities and the differences between the resonant and anti-resonant behaviour of metallic and dielectric structures are clear.

6.1 Metallic resonance probe

This probe consisting of a metal body with a DC bias and a RF signal superimposed on it has been introduced in 1960 by Takayama, Ikegami and Miyazaki[171]. Their measurements indicate that the incremental DC component Δi_{DC} is a function of frequency ω and that a marked resonance at $\omega = \omega_R$ is observed (see figure 6.3). They interpreted this ω_R as being equal to the local electron plasma frequency ω_e and this view was supported by the theoretical work of Ichikawa and Ikegami[172,173]. This interpretation was, however, challenged at Garching in 1963 where it was shown that, if one takes into account the ionic sheath present between a plane probe and the plasma, the resonance occurs in fact at $\omega_R = \theta\omega_e$ $(0 < \theta < 1)$, where θ is a function of the sheath (Mayer[174], Peter, Müller, Rabben[175,176], Wimmel[177]). The above result was qualitatively interpreted on the basis of the theory of the plasma slab-condenser system studied in chapter 2. This is done by characterizing the ionic sheath by the vacuum permittivity ϵ_0 but an unsolved problem remains, of course, the determination of the equivalent condenser. Such a model is even more directly connected to the experiment of Levitskii and Shashurin[178] in which they examine the high-frequency signal transmitted between two probes immersed in a plasma and in which they show that the resonance frequency depends on the DC bias, i.e. on the thickness of the sheaths; they also find that there exists an anti-resonance (no transmission of signal) for $\omega = \omega_e$ as is predicted theoretically by the plasma slab-condenser system. The high-frequency properties of a plasma sheath were studied by Pavkovitch and Kino[179] by direct solution of Boltzmann-Vlasov's equation and they show that, in agreement with the fluid theory results of chapter 2, the RF electric field in the sheath and in the plasma are 180° out of phase. Applying these last results to the metallic resonance probe, after using a correction factor to take the spherical geometry into account, Harp and Crawford[180,181] find that the resonance frequency is given by

$$\omega_R = \omega_e \sqrt{\frac{s}{R+s}} \qquad (6\text{-}1)$$

where R is the radius of the probe and s the equivalent thickness of the sheath. A general analysis has been performed by Messiaen[182,183] in several geometries by considering the problem of the metallic probe as a boundary value problem; he finds a resonance value in the spherical case which agrees with that given by equation (6-1). Comprehensive experiments have been carried out (Harp, Crawford[180,181], Dote, Ichimiya[184], Peter[185]). Further experiments (Lepechinsky, Messiaen, Rolland[187,188]) show that there is very good agreement between theory and experiment, but that the resonance value ω_R can only be used to measure the plasma density accurately if the detailed structure of the sheath is also known. The anti-resonance occurring at $\omega = \omega_e$, which is difficult to observe in a precise way, can be determined more easily by modulating the plasma sheath with a low-frequency signal (50 Hz): all the curves corresponding to the probe response must go through the same minimum defined by $\omega = \omega_e$ (Peter[185,186]). Finally, we mention that a good review article giving many details has been written by Taillet[189].

Metallic Resonance Probe

A modification of this probe has been considered by Smith and Bekefi[191]: they drive two halves of a metal sphere with a RF generator. Detailed measurements[192] show quite good agreement with theory and more details are given in reference 19.

High-frequency analysis

We now give the theoretical description of the probe following reference 182. We consider a metal body immersed in an infinite plasma and having

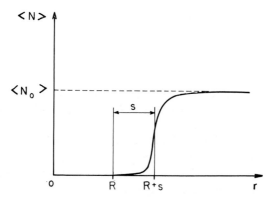

Figure 6.1 Electron density distribution $\langle N \rangle$ in the neighbourhood of the probe as a function of r; R is the 'radius' of the probe. The equivalent thickness of the sheath is s

a DC potential $V_0 < 0$ with respect to the plasma potential at infinity. This potential V_0 and the geometrical shape of the probe determine in the neighbourhood of the probe an electron density distribution since an ion sheath is created which separates the probe from the unperturbed plasma which we assume of uniform density $\langle N_0 \rangle$. Figure 6.1 gives such a typical electron density distribution where r is a coordinate orthogonal to the probe boundary.

A high-frequency signal of amplitude $\Phi_0(|\Phi_0| \ll |V_0|)$ is applied to the probe and we only consider the case for which the frequency $\omega/2\pi$ is small enough in order that the wavelength in vacuum λ_0 be such that $\lambda_0 \gg R + s$. We then can use the quasi-static approximation (see sections 1.3 and 2.1a) and set $\mathbf{E} = -\nabla\phi$. We further analyse the plasma in the cold plasma approximation and it is then described by its equivalent permittivity ϵ_p. The ionic sheath is represented by an equivalent vacuum layer of permittivity ϵ_0 and of thickness s; we have already seen in section 2.3 that this is a very good approximation. The high-frequency problem of the

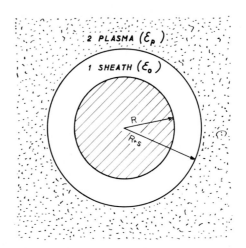

Figure 6.2 Spherical probe of radius R surrounded by a plasma sheath of thickness s and a uniform plasma of permittivity ϵ_p

probe is now cast in a form wholly analogous to that of chapter 2 and we have a boundary value problem of the type studied in chapters 3 and 4.

For the spherical probe of figure 6.2, the spherically symmetric solutions of Laplace's equation, having a correct behaviour in medium 1 and 2 respectively, are

$$\Phi_1 = A + B/r \quad (6\text{-}2)$$

$$\Phi_2 = C/r \quad (6\text{-}3)$$

The boundary conditions are:

(i) Equality of Φ and the applied RF potential Φ_0 ar $r = R$

$$A + B/R = \Phi_0 \quad (6\text{-}4)$$

(ii) Continuity of the potential Φ at $r = R + s$

$$A + B/(R+s) = C/(R+s) \quad (6\text{-}5)$$

(iii) Continuity of the normal component of the displacement vector at $r = R + s$.

$$-\epsilon_0 B/(R+s)^2 = -\epsilon_p C/(R+s)^2 \quad (6\text{-}6)$$

We compute the high frequency current density J_{HF} on the probe

$$J_{\text{HF}} = \epsilon_0 \left(\frac{\partial E_r}{\partial t}\right)_{r=R} = i\omega\epsilon_0 \frac{B}{R^2} = \frac{i\omega\epsilon_0\epsilon_p(R+s)\Phi_0}{R(s\epsilon_p + R\epsilon_0)} \quad (6\text{-}7)$$

Metallic Resonance Probe

If there are no collisions ($\nu/\omega = 0$), we have a resonance ($J_{HF} = \infty$) at

$$\omega_R = \omega_e \sqrt{\frac{s}{R+s}} \tag{6-8}$$

and an anti-resonance ($J_{HF} = 0$) at

$$\omega_{AR} = \omega_e \tag{6-9}$$

For small collisions frequencies ($\nu/\omega \ll 1$), the above values are approximately those for which $|J_{HF}|$ is maximum and minimum respectively.

Direct-current analysis of the probe

The variation of the DC probe current can be calculated by using the method of Boschi and Magistrelli[190] which neglects the transit time of the electrons in the sheath. Assuming that the relevant high-frequency difference of potential $\Delta\Phi_{HF}$ is that between the probe and the sheath-plasma interface, we have

$$\Delta\Phi_{HF} = -\int_R^{R+s} E_r \, dr = \frac{s\epsilon_p \Phi_0}{s\epsilon_p + R\epsilon_0} \tag{6-10}$$

The DC current density J_{DC} is given by

$$J_{DC} = J_{ODC} \frac{\omega}{2\pi} \int_0^{2\pi/\omega} e^{e|\Delta\Phi_{HF}|\sin\omega t/KT_e} \, dt = J_{ODC} I_0(e|\Delta\Phi_{HF}|/KT_e) \tag{6-11}$$

where J_{ODC} is the DC current density when V_0 is the probe potential, i.e. in the absence of a high-frequency signal and where I_0 is the modified Bessel function of the first kind of order zero. When $e|\Delta\Phi_{HF}|/KT_e \ll 1$, the incremental DC component $\Delta J_{DC} = J_{DC} - J_{ODC}$ of the current density is then

$$\Delta J_{DC} \simeq \frac{1}{4} \left(\frac{e|\Delta\Phi_{HF}|}{KT_e}\right)^2 J_{ODC} \tag{6-12}$$

ΔJ_{DC} has thus practically the same maximum (6-8) and the same minimum (6-9) as J_{HF}. This incremental DC component results of course from the fact that the sheath characteristic $J = f(V)$ is non-linear.

Discussion

A typical plot of $|J_{HF}/\omega|$ as a function of $\Omega^{-2} = (\omega/\omega_e)^2$ is given in figure 6.3. The reader will note the complete similarity with the curve giving J_{HF} versus $\Omega^2 = (\omega_e/\omega)^2$—and not Ω^{-2}— for the plasma slab-condenser system in figure 2.8. The reason why this plasma slab-condenser system furnishes such a good heuristic basis for the understanding of the metallic resonance probe is clear: the resonance is a series resonance and results from the balance of the positive capacitance of the sheath and the negative capacitance of the plasma when $\epsilon_p < 0$, i.e. $\omega < \omega_e$, and ω_R is thus

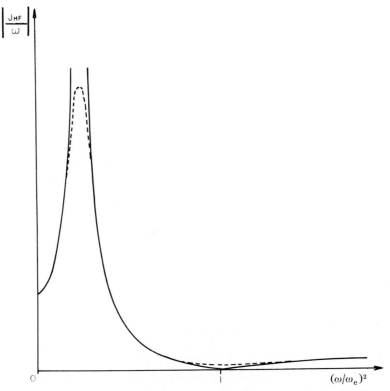

Figure 6.3 $|J_{\mathrm{HF}}/\omega|$ versus $\Omega^{-2} = (\omega/\omega_e)^2$ for a typical metallic resonance probe with no collisions and for small collision frequencies. Where ———: $\nu/\omega = 0$ and ·····: $\nu/\omega \ll 1$

equal to $\omega_e \theta$ with $0 < \theta < 1$; the anti-resonance occurs when $\epsilon_p = 0$ (if $\nu/\omega = 0$) because the continuity of the normal component of the electric displacement then imposes that the electric field in the sheath, and thus J_{HF} of the probe, be zero. Further details on this point will be given in section 6.4.

If one takes into account that $T_e \neq 0$, one finds, in accordance with the analysis of chapters 2 and 5, a set of supplementary resonances which can be observed if ν/ω is sufficiently small but which play no important role here. A numerical experiment simulating the resonance probe has been

Metallic Resonance Probe

performed by Hellberg[196]. The trajectories of a large number of individual electrons are followed in a self-consistent calculation in plane geometry. The results obtained largely bear out, as expected by what has been seen in the previous chapters, those obtained by the less sophisticated but realistic cold plasma approach.

Detailed comparison between theory and experiments[187,188] shows that:
(1) There is fair agreement between the values of ω_e deduced from the anti-resonance and those obtained from the electron branch of the Langmuir characteristic of the probe (see figure 6.4).
(2) There is a good agreement between the values of the sheath thickness s obtained from the ion-sheath theoretical curves of Chen[193] and those deduced from ω_R (see figure 6.4).
(3) It is verified that $|\Delta J_{DC}| \propto |\Phi_0|^2$ whereas $|J_{HF}| \propto |\Phi_0|$.
(4) It can be shown[194] that the ratio $R = |\Delta J_{DC(\omega=\omega_R)}|/|\Delta J_{DC(\omega=0)}|$ leads to a simple method for the determination of the collision frequency ν and it is found that the values of ν deduced from the ratio R agree with those derived from direct measurements of the electron mean free path[195]*.

The general conclusion is that good agreement exists between theory and experiments but that resonance probe diagnostics require careful interpretation of the results and a good knowledge of the sheath. The determination of the plasma density by the experimentally ill-defined anti-resonance can be somewhat improved by the technique of low frequency modulation of the sheath indicated above[185,186].

A point to be mentioned is the behaviour of the probe in a static magnetic induction $\mathbf{B}_0$. The application of the method described in chapter 4 leads immediately to the fact that the different resonances should couple to each other. It should, however, be stressed that the validity of such a theoretical result is extremely limited. Indeed, we know that very small magnetic fields (a few Gauss) deform the sheath anisotropically in a significant way and it is illusory then to make a quantitative computation involving the sheath thickness in the absence of a magnetic field. Any realistic theory must necessarily be rather complicated since it should include the influence of the steady magnetic field on the sheath.

Problem: Metallic resonance probe coated with a dielectric layer (from reference 198). Show that the high-frequency current density collected by the probe is:

$$J_{HF} = \frac{i\omega\epsilon_1\Phi_0}{R_1} \times \frac{R_2(R_2+s)\epsilon_0\epsilon_p}{R_1R_2\epsilon_1 + [(R_2+s)(R_2-R_1)\epsilon_0 + sR_1\epsilon_1]\epsilon_p}$$

where Φ_0 is the amplitude of the applied RF voltage (see figure 6.5).

* This is only correct if there is no appreciable damping due to phase mixing (collisionless damping).

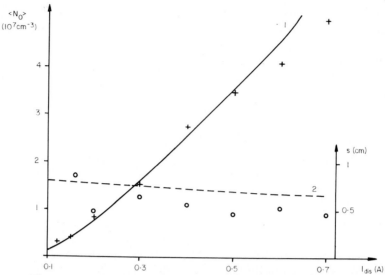

Figure 6.4 Electron plasma density $\langle N_0 \rangle$ and sheath thickness s versus discharge current in a mercury vapour plasma. Curve 1 gives the Langmuir probe data and the crosses indicate the corresponding RF probe data. Curve 2 gives the sheath deduced from ion-sheath theoretical computations and the dots indicate the corresponding RF probe data (after reference 188)

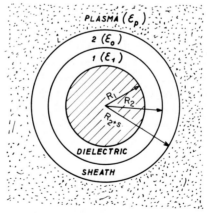

Figure 6.5 Metallic resonance probe coated with a dielectric layer

6.2 High-frequency dielectric resonance probe

The resonance frequency of the metallic resonance probe studied in the previous section is very sensitive to the exact value adopted for the sheath thickness. We now describe a high-frequency dielectric resonance probe displaying both a resonance and an anti-resonance which are, to a large extent, insensitive to this sheath[199,200]. This probe is constituted by two

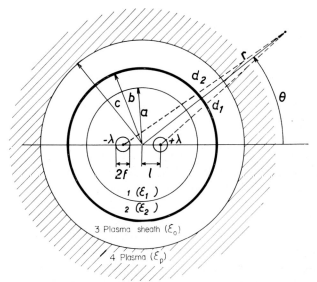

Figure 6.6 The two-wire dielectric probe, placed in dielectric medium 1, is surrounded by the plasma sheath (3) and the plasma (4). Medium 2 is a further possible dielectric layer

wires placed inside a dielectric cylinder isolating them from the neighbouring plasma (figure 6.6).

As is well known (e.g. sections 1.4 and 3.1), a uniform plasma column of radius a placed in vacuum has a geometrical main resonance frequency $\omega_e/\sqrt{2}$ independent of a in the quasistatic approximation ($a \ll \lambda_0$; λ_0: vacuum wavelength). It was noted[21] that the reciprocal system consisting of a cylinder of vacuum of radius a embedded in an infinite uniform plasma possesses the same eigenfrequency. This property can, in fact, be stated in a slightly more general way:

Problem: Show that the eigenfrequency of the system consisting of two concentric cylinders (of radii a and b, with permittivity ϵ_1 and ϵ_2, respectively, (ϵ_1 and $\epsilon_2 > 0$) embedded in an infinite uniform plasma remains the same if the innermost medium characterized by ϵ_1 and the outer plasma are interchanged. Show that this is only true in cylindrical geometry and that in spherical geometry, for example, there exists a resonance frequency ($<\omega_e$) in both cases, but that these frequencies are not equal.

In this context, it should be mentioned that a further detailed analysis of the resonant frequencies of cavities in a magnetized plasma has been performed by Budden[197].

The fact that a cylinder of vacuum embedded in an infinite plasma has a resonance frequency $\omega_e/\sqrt{2}$ independent of a if $a \ll \lambda_0$ led to a proposal of exciting this eigenfrequency through a two-wire probe placed in the innermost medium[198] because this property heuristically suggests that the resonance frequency of such a system will only slightly depend on the sheath which predominantly behaves as a vacuum layer. In the remainder of this section we now follow rather closely the papers of Messiaen and Vandenplas devoted to the subject[199,200].

a. Theory

The dielectric two-wire probe is described in figure 6.6. As in the previous section, the sheath is described as being a layer of vacuum and the plasma in 4 is uniform. This approximation of the steep density variation in the neighbourhood of the dielectric is even more warranted as it will be shown that the sheath will have but a small influence on the resonance and the anti-resonance.

A high-frequency voltage is applied between the wires. The quasi-static approximation is adopted since $c \ll \lambda_0$. The problem is solved in the cylindrical coordinate system (r, θ, z) indicated in figure 6.6 and is treated in the following fashion. When $f/l \ll 1$, the 'applied' potentials ϕ_{a+} and ϕ_{a-} can be considered as resulting from two given linear charge densities $+\lambda$ and $-\lambda$, placed at $r = l, \theta = 0°$ and $r = l, \theta = 180°$, respectively, which induce the potentials ϕ_1, ϕ_2, ϕ_3 and ϕ_4, in the different media 1, 2, 3 and 4. The potential ϕ_a is thus the only one that would exist if medium 1 were infinite and is given by

$$\phi_a = \phi_{a+} + \phi_{a-} = -(\lambda/2\pi\epsilon_1)(\log_e d_1 - \log_e d_2) \quad (6\text{-}13)$$

In cylindrical coordinates one has:

$$\phi_a = \begin{cases} \dfrac{\lambda}{\pi\epsilon_1} \sum_{n=1,3,5\cdots} \dfrac{1}{n} \left(\dfrac{l}{r}\right)^n \cos n\theta; & r > l \\ \dfrac{\lambda}{\pi\epsilon_1} \sum_{n=1,3,5\cdots} \dfrac{1}{n} \left(\dfrac{r}{l}\right)^n \cos n\theta; & r < l \end{cases} \quad (6\text{-}14)$$

High-Frequency Dielectric Resonance Probe

Only the odd multi-poles characterized by n odd are excited. The induced potentials are solutions of the Laplace equation and take the form:

$$\phi_1 = \sum_{n=1,3,\ldots} A_n r^n \cos n\theta$$

$$\phi_2 = \sum_{n=1,3,\ldots} [B_n r^n + C_n r^{-n}] \cos n\theta$$

$$\phi_3 = \sum_{n=1,3,\ldots} [D_n r^n + G_n r^{-n}] \cos n\theta \qquad (6\text{-}15)$$

$$\phi_4 = \sum_{n=1,3,\ldots} L_n r^{-n} \cos n\theta$$

The boundary conditions at each surface of discontinuity are given by the continuity of the potential and the continuity of the normal components of the electric displacement. This leads to six equations to determine the six unknowns $A_n, \ldots, L_n$. The value of A_n is thus obtained:

$$A_n = \left(\frac{l}{a^2}\right)^n \frac{\lambda}{n\pi\epsilon_1} \psi_n \qquad (6\text{-}16)$$

$$\psi_n = \frac{\begin{array}{c}-\epsilon_p[\epsilon_0(\epsilon_2 + \epsilon_1 x_n) + y_n \epsilon_2(\epsilon_1 + \epsilon_2 x_n)] \\ + \epsilon_0[\epsilon_2(\epsilon_2 x_n + \epsilon_1) + y_n \epsilon_0(\epsilon_2 + \epsilon_1 x_n)]\end{array}}{\begin{array}{c}\epsilon_p[\epsilon_0(\epsilon_2 - \epsilon_1 x_n) - y_n \epsilon_2(\epsilon_1 - \epsilon_2 x_n)] \\ - \epsilon_0[\epsilon_2(\epsilon_2 x_n - \epsilon_1) + y_n \epsilon_0(\epsilon_2 - \epsilon_1 x_n)]\end{array}}$$

$$x_n = (a^{2n} - b^{2n})/(a^{2n} + b^{2n}), \qquad y_n = (b^{2n} - c^{2n})/(b^{2n} + c^{2n})$$

As $f/l \ll 1$, the potential of each wire is given by $\phi_a + \phi_1$ at $r = l$, $\theta = 0°$ and $r = l$, $\theta = 180°$ respectively, and is noted ϕ_0 or $-\phi_0$. The capacity per unit of length between the two wires is given by

$$C = \frac{\lambda}{2\phi_0} = \frac{1}{2}\left[\frac{\phi_a}{\lambda} + \frac{\phi_1}{\lambda}\right]^{-1}_{r=l,\theta=0} \qquad (6\text{-}17)$$

The expression for $C_a = \lambda/2\phi_a$ is well known:

$$C_a = \pi\epsilon_1/(\cosh^{-1} l/f)$$

The impedance Z per unit of length between the two wires is thus

$$Z = \frac{i}{\omega C} = \frac{i}{\omega C_a} + \frac{2i}{\omega \pi \epsilon_1} \sum_{n=1,3,5,\ldots} \frac{1}{n}\left(\frac{l}{a}\right)^{2n} \psi_n \qquad (6\text{-}18)$$

If we take $l/a \ll 1$, the dipolar term ($n = 1$) is the only important one and

$$Z \simeq \frac{i}{\omega C_a} + \frac{2i}{\omega \pi \epsilon_1}\left(\frac{l}{a}\right)^2 \psi_1 = \frac{2i}{\omega \pi \epsilon_1}\left(\frac{l}{a}\right)^2 (K + \psi_1) \qquad (6\text{-}19)$$

with

$$K = \tfrac{1}{2}(a/l)^2 \cosh^{-1} l/f$$

It is very important to note that $K \gg 1$.

Discussion of Z

If we neglect collisions, we have the well-known expression $\epsilon_p = \epsilon_0(1 - \omega_e^2/\omega^2)$. The *anti-resonance* ($Z = \infty$) is obtained from equation (6-19) and occurs for the following value of ϵ_p:

$$\frac{\epsilon_p}{\epsilon_0} = -F \frac{1 - y_1/F}{1 - y_1 F} = -S; \quad F = \frac{\epsilon_2(\epsilon_1 - \epsilon_2 x_1)}{\epsilon_0(\epsilon_2 - \epsilon_1 x_1)} \quad (6\text{-}20)$$

Case 1: If the dielectric surrounding the two wires is characterized by $\epsilon_1 = \epsilon_2 = \epsilon_0$, F is equal to 1 and $\epsilon_p = -\epsilon_0$. The anti-resonance frequency is then given by $\omega_e/\sqrt{2}$; this eigenfrequency is naturally the same as that of the model which is the basis of the present study and it is, of course, independent of the sheath thickness.

Case 2: It must be noticed that the value of ϵ_p given by equation (6-20) is only a function of the sheath thickness through y_1 and that it becomes practically independent of the sheath when

$$|y_1| = \left| \frac{b^2 - c^2}{b^2 + c^2} \right| \ll 1$$

i.e. when the sheath thickness is small compared to the radius of the probe.

We now discuss the *resonance* ($Z = 0$) which is more of practical importance since it is more easily observed experimentally than the anti-resonance. The resonance occurs for

$$\frac{\epsilon_p}{\epsilon_0} = -H \frac{1 - y_1/H}{1 - y_1 H} = -T; \quad H = \frac{K\epsilon_2(\epsilon_1 - \epsilon_2 x_1) + \epsilon_2(\epsilon_1 + \epsilon_2 x_1)}{K\epsilon_0(\epsilon_2 - \epsilon_1 x_1) - \epsilon_0(\epsilon_2 + \epsilon_1 x_1)}$$

(6-21)

Case 1: If $\epsilon_1 = \epsilon_2 = \epsilon_0$, F is equal to 1 and $H = (K+1)/(K-1)$. As $K \gg 1$, H is then $\simeq 1$. By setting $a = b = c$, we obtain the resonance frequency:

$$\omega_{\text{Res}} = \omega_e \sqrt{\frac{1}{2} - \frac{1}{2K_{a=c}}} \quad (6\text{-}22)$$

We notice that the resonance and anti-resonance frequencies are close to each other and the resonance will thus only be easily detected if $\nu/\omega \ll 1$ (ν is the 'average' collision frequency). This is the case of the ionosphere; the two-wire probe, provided $H \simeq 1$, should then prove a very good diagnostic tool giving a value of plasma density which depends very little on the structure of the plasma sheath. This is not, of course, the case with the metallic resonance probe. A value of H near 1 can be obtained by using dielectric foams which have a dielectric constant lying near 1 (e.g. 1·03 to 1·5); we then have $\epsilon_1 = \epsilon_2 \simeq \epsilon_0$.

Case 1 bis: Another way of easily obtaining $H \simeq 1$ is the following. The two wires are placed in vacuum or air and are surrounded by a thin dielectric cylinder $[(b - a)/a \ll 1]$. Then $x_1 \ll 1$ and $H \simeq (K + 1)/(K - 1) \simeq 1$. Again such a probe can easily be built on a satellite.

Case 2: It must also be noted that the resonance value of ϵ_p given by equation (6-21) depends only on the sheath thickness through y_1. This resonance value will become practically independent of the sheath when the

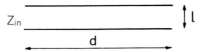

Figure 6.7   The two-wire probe is open-circuited to the right and acts as a transmission line as $d = 0(\lambda_0)$

sheath thickness is small compared to the radius of the probe. As the sheath thickness $s = c - b$ is of the order of the Debye radius r_D, one has the following criterion:

$$r_D/b \ll H/(H^2 - 1). \tag{6-23}$$

To have an idea of the precision of the method in this case, we study an example. The density variation of the plasma near the probe is approximated by a constant density plasma and a sheath of thickness $3r_D$. If $H = 2$, $T_e = 10^4 \,°K$, $b = 0·5$ cm, the relative error on the measure of the density is smaller than 10 per cent if the electron plasma density N_e is greater than 4×10^8 cm^{-3}.

High density limit

If the density becomes high, ω_e and ω_R become high and the criterion for the quasistatic approximation ($a \ll \lambda_0$) involves the miniaturization of the probe. If both transverse dimensions of probe and length of wires are small, we have case 2, as the Debye length and sheath are small. It proves easier to miniaturize the transverse dimensions and to have a length of wire of the order of λ_0. We then have a transmission line as described in figure 6.7. It has a characteristic impedance of $Z_0 = (L/C)^{\frac{1}{2}}$ with $L = (\mu_0/\pi) \cosh^{-1} l/f$, C given by equation (6-19) and a propagation constant $k = \omega(LC)^{\frac{1}{2}}$. The impedance Z_{in} seen by the measurement setup is

$$Z_{in} = Z_0 \frac{Z_T + iZ_0 \tan kd}{Z_0 + iZ_T \tan kd}; \quad Z_T: \text{impedance of terminal} \quad (6-24)$$

As we have an open-circuited line, $Z_T = \infty$ and equation (6-24) becomes
$$Z_{in} = -iZ_0 \cot kd. \tag{6-25}$$
The anti-resonance is given by $C = 0$ and the impedance Z_{in} is then infinite for any value of d. The resonance is given by $C = \infty$ and by applying the rule of l'Hospital, one sees that $Z_{in} = 0$ for any value of d. The resonance can thus be easily observed by using this transmission line setup.

A typical example in this high density limit is the following. If $H = 2$, $b = 0.15$ cm, a density N_e of 10^{13} cm^{-3} can be measured with a relative error of 10 per cent if $T_e < 5 \times 10^7$ °K.

Influence of a steady magnetic induction **B**$_0$

As the sheath only slightly affects the resonance frequency, we must expect that the influence of a steady magnetic induction **B**$_0$ on this resonance can be calculated in a more trustworthy way than for the spherical metallic probe where a serious difficulty arises from the asymmetrical influence of the magnetic field on the sheath. Moreover, when the magnetic field is parallel to the cylindrical dielectric probe, it affects the sheath in a symmetrical fashion and will only change its thickness.

To calculate the influence of the magnetic field, we need only introduce the tensorial permittivity $\boldsymbol{\epsilon}_p$ given by equation (1-23) instead of the scalar permittivity ϵ_p. The normal component of the electric displacement vector **D**$_4$ in the plasma then becomes

$$D_{4r} = -\left(\epsilon_{rr}\frac{\partial \phi_4}{\partial r} + \epsilon_{r\theta}\frac{1}{r}\frac{\partial \phi_4}{\partial \theta}\right) \tag{6-26}$$

Equation (6-26) implies a coupling between the solutions in $\cos n\theta$ and those in $\sin n\theta$ when the boundary conditions between the different media are expressed. Whereas the applied potential (6-14) is only a function of $\cos n\theta$, the induced fields in the four different media will also contain $\sin n\theta$ terms in order to satisfy the boundary conditions. There is therefore twice as many equations to determine the impedance between the two wires as in the absence of a magnetic field.

Retaining only the dipolar term ($n = 1$), we now have two anti-resonances and two resonances. The anti-resonances ($Z = \infty$) are given by

$$\frac{\omega_e^2}{\omega^2} = (1 + S)(1 \pm \beta); \quad \beta = \omega_c/\omega = eB_0/m\omega \tag{6-27}$$

and the resonances ($Z = 0$) by

$$\frac{\omega_e^2}{\omega^2} = 1 + \frac{T+S}{2} \pm \frac{T-S}{2}\sqrt{1 + 4\beta^2\frac{(1+T)(1+S)}{(T-S)^2}} \tag{6-28}$$

where S and T are defined in equations (6-20) and (6-21) respectively.

High-Frequency Dielectric Resonance Probe

If $K \gg 1$, the resonances are rather near to the anti-resonances as in the absence of a magnetic field ($\beta = 0$). The cases for which the resonance frequencies depend only slightly on the sheath thickness are identical to those studied without magnetic field. This slight dependence on the sheath thickness indicates *a posteriori* that the probe should give good results for moderate values of the magnetic field*.

b. Experiments

No magnetic field

Experiments were performed without magnetic field in a situation corresponding to *Case 2*. The plasma is generated in a low pressure Hg discharge tube. The resonances, given by Z of equation (6-19) equal to zero, can be observed in transmission or in reflexion. We note that the transmission experiments are somewhat more unwieldy since the probe must traverse the vessel in which the plasma is created (figure 6.8).

Figure 6.10a shows the resonance exhibited by the high-frequency voltage in a reflexion experiment when the discharge current I_{dis}, which is

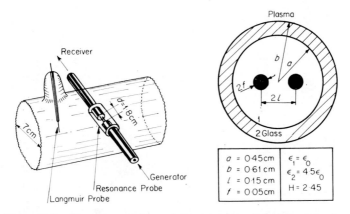

Figure 6.8 The plasma tube and the two-wire dielectric probe together with a section of the probe and the relevant dimensions

---

* It should be mentioned here that a case intermediate between the usual metallic resonance probe and the dielectric wire probe has been recently studied (Crawford, Harp and Martei[220]), namely a metallic parallel wire configuration immersed in a plasma. The magnetic field is placed parallel to the wires and thus affects the sheath symmetrically. The probe is studied at values of the magnetic field where the cyclotron harmonics play a role. As seen in section 5.3, this means that the theoretical description of the plasma must necessarily be that of a hot one.

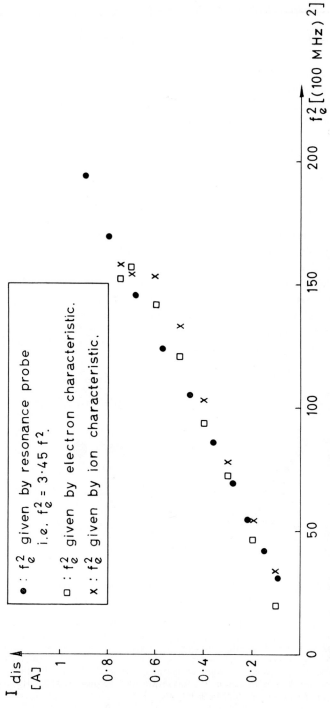

Figure 6.9 Discharge current versus electron plasma density, shown as $f_e^2 = (\omega_e/2\pi)^2$, measured through the ion and electron characteristics of a Langmuir probe. Same discharge current as a function of $f_e^2 = 3 \cdot 45 f^2$ (theory of dielectric probe, no sheath)

proportional to the electron density $\langle N \rangle$, is varied. Typical results are listed in figure 6.9. The discharge current is shown as a function of ω_e^2 assuming that the theoretical formula without sheath ($y_1 = 0$), in the experimental situation considered,

$$\omega_e^2 = 3 \cdot 45 \omega_R^2 \qquad (6\text{-}29)$$

is correct.

The electron densities were measured independently for each discharge current by the method of the electron characteristic and also by the ion characteristic (using the numeral calculations by Chen[193]) of a thin cylindrical Langmuir probe. These values of I_{dis} vs ω_e^2 are also shown in figure 6.9. Note that there is no dispersion of the resonance results (successive measurements at a given frequency lead to a set of experimental points on

Table 6.1 $\langle N \rangle$ in cm^{-3}

Without sheath	With sheath = $3\lambda_D$
3·85 10^9	3·45 10^9
2·41 10^{10}	2·36 10^{10}

figure 6.9 which practically coincide). This is, of course, not true for the results obtained with the ion and electron characteristics where a dispersion of approximated 15 to 20 per cent exists which is inherent in the method.

Although we have a rather high value of H ($H = 2 \cdot 45$) in this experiment, the influence of the sheath remains rather small. Table 6.1 gives the density corresponding to the theoretical formula without sheath and the corrected value with a sheath thickness $|b - c| = 3\lambda_D$ for the lowest and the highest density considered. This unfavourable value of H can be easily lowered in a more appropriate setup and the already small influence of the sheath will then be further reduced.

At high frequency, the classical temperature or secondary resonances of chapter 5 have also been observed with this probe.

One facility when using the metallic probe, however, lies in the fact that the resonance can be observed in the direct current resulting from the non-linear characteristic of the probe. A non-linear element can also be introduced in the present dielectric probe by placing a small silicon diode between the two wires. This diode can be suitably biased in order that the operating domain be situated in the most non-linear part of the diode characteristic. One obtains a curve giving I_{DC} as a function of the electron plasma density $\langle N \rangle$. This curve displays a marked resonance at the same density for which the high-frequency signal has a resonance.

A further appraisal of this new dielectric probe has been done by F. Smith and T. Johnston[201]. They have compared the densities obtained with the probe to those obtained by the double Langmuir probe technique. They find very good agreement and indicate that when a slight discrepancy arises at one end of their observation domain, it is very probably ascribable to the double probe method and not to the dielectric probe.

With magnetic field

The splitting of the resonance peak in an axial steady magnetic induction $\mathbf{B}_0$ which is predicted by equation (6-28) is all the more easily observable that the ratio v/ω is small (v: collision frequency). Figures 6.10a and 6.10b show typical oscillograms of the reflected signal versus $I_{dis} \propto \langle N \rangle$, with and

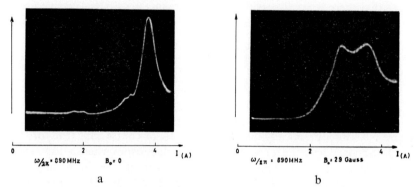

Figure 6.10 (a) Reflected signal as a function of I_{dis} ($\propto \langle N \rangle$) when $B_0 = 0$. Vertical gain: 1.
(b) Reflected signal as a function of I_{dis} ($\propto \langle N \rangle$) with an applied axial B_0. Vertical gain: 3

without axial $\mathbf{B}_0$, which clearly exhibit this splitting. The electron plasma densities measured, for given values of $\mathbf{B}_0$ and of I_{dis}, by one or the other resonance frequency must be identical since the density $\langle N \rangle$ depends only on I_{dis} and $\mathbf{B}_0$. This has been verified for low magnetic fields and the results are given in Table 6.2. With higher magnetic fields, β increases and the agreement between theory and experiments is not as good. The situation is in fact the same as for the resonance frequencies of a plasma cylinder in an axial magnetic field (section 4.2). Attempts to include the non-uniformity of the plasma, the effect of the electron temperature or the collision effects have not yet been able to explain this discrepancy[202].

When $\mathbf{B}_0$ is perpendicular to the axis of the probe, the behaviour is similar to that observed in the same situation with the plasma column placed in vacuum (section 4.3). There is an enhancement of the cold multipolar resonances ($n > 1$) and of the temperature resonances; the analysis of section 4.3 indicates that the detailed interpretation of these results is not a straightforward matter. However, we can infer from this analysis that for rather low values of $\mathbf{B_0}$, the effect on the *position* of the main resonance will be negligible and that formula (6-21), valid for $\mathbf{B_0} = 0$, can still be applied. It is experimentally verified that when $\beta = 0.1$, the values of $\langle N \rangle$ obtained by placing the probe perpendicular to $\mathbf{B_0}$ and applying equation (6-21) or placing it parallel to $\mathbf{B_0}$ and applying equation (6-28) are nearly identical.

Table 6.2 Comparative measurements of electron plasma density $\langle N \rangle$ from the two resonance frequencies f_1 and f_2 in an axial $\mathbf{B_0}$

I_{dis} (A)	B_0 (Gauss)	Resonance Frequencies (MHz)		Calculated Density (cm^{-3}) (equation (6-28) with $S = 2.11$ and $T = 2.35$)	
		f_1	f_2	From f_1	From f_2
1.55	33.9	700	600	1.73×10^{10}	1.65×10^{10}
2.15	30	800	700	2.27×10^{10}	2.22×10^{10}
2.55	40.5	948	850	3.32×10^{10}	3.14×10^{10}

Problem: Show that in the presence of an axial steady magnetic induction B_0, the resonance frequencies of a cylinder of vacuum embedded in an infinite plasma are, in the quasi-static approximation, identical to these of the reciprocal system, i.e.

$$\frac{\omega_e^2}{\omega^2} = 2\left(1 \pm \frac{\omega_c}{\omega}\right) \tag{6-30}$$

(from reference 197).

It is clear that the simultaneous use of the dielectric resonance probe and the metallic one provides a rapid and precise measurement of both the density and the electron temperature of the plasma.

6.3 Enhanced radiation from an antenna coated with dielectric and plasma

An antenna surrounded by a plasma layer exhibits very strongly enhanced radiation at certain plasma densities for fixed operating frequency and it is shown both experimentally and theoretically that the plasma sheath surrounding the aerial plays an essential role in this resonant behaviour (Messiaen, Vandenplas[203,204]). That the role of the plasma sheath proves to be an important feature is certainly no surprise after having studied the plasma slab-condenser system (chapter 2) and the metallic resonance probe (section 6.1). In fact, the metallic probe is also an antenna placed in a practically infinite plasma but since it resonates at $\omega_R = \omega_e \theta$ ($0 < \theta < 1$), the wave that it launches in the plasma is evanescent because $k^2 = \omega^2 \mu_0 \epsilon_p$ is then negative. The striking difference between the antenna to be described and the probe is that the thickness of the plasma layer is small compared to the vacuum wavelength λ_0 and that the plasma therefore has no screening effect when $\omega < \omega_e$. At $\omega < \omega_e$, $k^2 < 0$ and this thin plasma layer has an inductive behaviour: it can form a resonant system with the sheath and also a resonant system with the vacuum outside; we shall see that the resonance frequencies of these two systems are very loosely coupled.

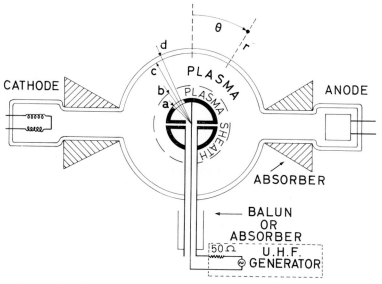

Figure 6.11 Discharge tube with spherical slotted antenna, plasma sheath and plasma, together with feeding setup

A typical antenna for which theory and experiments can be easily compared consists of two isolated metallic hemispheres of radius a fed by a generator through a coaxial cable terminated by a balun or coated with an absorber (figure 6.11). The plasma is created in a low-pressure mercury-discharge tube which has a spherical region between the glass wall of the discharge tube and the antenna. The thickness $s = b - a$ of the plasma sheath is determined by the DC potential of the antenna. Absorbing screens are placed around the cylindrical part of the discharge tube, in order to minimize spurious effects.

a. Theory

Theoretically, the plasma is assumed uniform and described in the cold plasma limit by its equivalent permittivity ϵ_p. As previously seen in chapter 2 and in this chapter, the plasma sheath has a vacuum permittivity ϵ_0.

Enhanced Radiation from an Antenna

The hemispheres have an infinite conductivity; the upper lip of the antenna is at the RF potential $+\phi_0 e^{-i\omega t}$ and the lower lip at $-\phi_0 e^{-i\omega t}$. These are the usual assumptions made in the treatment of the split-spherical antenna radiating in vacuum (see reference (18)). We use a r, θ, ϕ spherical coordinate system as indicated in figure 6.11. By symmetry, we have $\partial/\partial\phi \equiv 0$. The problem is solved by developing in spherical functions the excitation and the solutions of Maxwell's equations in the different media considered and expressing, at the interfaces of these media, the continuity of the tangential components of the electric and magnetic fields. When $r \to \infty$, we must satisfy Sommerfeld's radiation condition. With the excitation considered and since there is no anisotropy in the plasma, only TM solutions ($H_r = 0$) are present.

The applied electric fields strength is such that $E_\theta = 0$ at $r = a$ except between the lips where

$$2\phi_0 = \lim_{\alpha \to 0} \int_{\pi/2+\alpha}^{\pi/2-\alpha} E_\theta a \, d\theta \qquad (6\text{-}31)$$

We therefore have

$$(E_\theta)_{r=a} = \frac{2\phi_0}{a} \delta(\cos\theta)$$

$$= \frac{2\phi_0}{a} \sum_{n=1,3,5,\ldots} (-1)^{(n-1)/2} \frac{2n+1}{2^{n+1}} \frac{(n-1)!}{\left(\frac{n-1}{2}\right)! \left(\frac{n+1}{2}\right)!} P_n^1(\cos\theta)$$

$$(6\text{-}32)$$

where $\delta(\cos\theta)$ is the Dirac delta function and $P_n^1(\cos\theta)$ is an associated Legendre polynomial of degree n and order 1. The TM solutions of Maxwell's equations having revolution symmetry have three components H_ϕ, E_θ and E_r. In media 1, 2 and 3, these are

$$H_{\phi i} = -i\omega\epsilon_i \sum_{n=1,3,\ldots} \{A_{in} j_n(k_i r) + B_{in} n_n(k_i r)\} P_n^1(\cos\theta) \qquad (6\text{-}33)$$

$$E_{\theta i} = -\frac{1}{r} \sum_{n=1,3,\ldots} \left\{ A_{in} \frac{d}{dr}[r j_n(k_i r)] + B_{in} \frac{d}{dr}[r n_n(k_i r)] \right\} P_n^1(\cos\theta) \qquad (6\text{-}34)$$

$$E_{ri} = \frac{1}{r} \sum_{n=1,3,\ldots} n(n+1)\{A_{in} j_n(k_i r) + B_{in} n_n(k_i r)\} P_n(\cos\theta) \qquad (6\text{-}35)$$

where $i = 1, 2, 3$ and $k_i^2 = \epsilon_i \mu_0 \omega^2$, $\epsilon_1 = \epsilon_0$, $\epsilon_2 = \epsilon_p$, $\epsilon_3 = \epsilon_g$. A_{in} and B_{in} are the constants to be computed by the boundary conditions while $j_n(k_i r)$ and $n_n(k_i r)$ are respectively the spherical Bessel functions of the first and second kind of order n. In the vacuum outside (medium 4), the electromagnetic field having a proper behaviour at infinity is

$$H_{\phi_4} = -i\omega\epsilon_0 \sum_{n=1,3,\ldots} F_n h_n^{(1)}(k_0 r) P_n^{1}(\cos\theta) \tag{6-36}$$

$$E_{\theta_4} = -\frac{1}{r} \sum_{n=1,3,\ldots} F_n \frac{d}{dr}[r h_n^{(1)}(k_0 r)] P_n^{1}(\cos\theta) \tag{6-37}$$

$$E_{r_4} = \frac{1}{r} \sum_{n=1,3,\ldots} n(n+1) F_n h_n^{(1)}(k_0 r) P_n(\cos\theta) \tag{6-38}$$

where $h_n^{(1)}(k_0 r)$ are the spherical Hankel functions of the first kind and $k_0^2 = \epsilon_0 \mu_0 \omega^2$.

The constants A_{in}, B_{in} and F_n are determined by the boundary conditions on H_ϕ and E_θ after taking into account that the total E_θ in medium 1 is given by the sum of the excitation field (6-32) and the induced field (6-34). We obtain in this way a system of seven transcendental equations to determine the amplitude F_n of the field radiated in the outer vacuum. Since $d \ll \lambda_0$, we use the asymptotic expressions of the different spherical functions for $|kr| \ll 1$, This leads to an algebraical expression for F_n:

$$F_n = i(-1)^{(n+1)/2} \frac{\phi_0(k_0 a)^{n+1}[(n-1)!]^2}{2(2n-1)!\left(\frac{n-1}{2}\right)!\left(\frac{n+1}{2}\right)!} \left(\frac{2n+1}{n}\right)^4 \frac{\epsilon_0 \epsilon_g \epsilon_p (bcd)^{2n+1}}{\Delta_n} \tag{6-39}$$

with

$$\Delta_n = Q_n\left[\epsilon_0 \frac{n+1}{n}(d^{2n+1} - c^{2n+1}) + \epsilon_g\left(\frac{n+1}{n}c^{2n+1} + d^{2n+1}\right)\right]$$

$$+ \epsilon_g P_n\left[\epsilon_0\left(c^{2n+1} + \frac{n+1}{n}d^{2n+1}\right) + \epsilon_g(d^{2n+1} - c^{2n+1})\right]$$

$$+ i\rho_n d^{2n+1}\left\{Q_n\left[\epsilon_0 \frac{n+1}{n}(d^{2n+1} - c^{2n+1})\right.\right.$$

$$\left.- \frac{n+1}{n}\epsilon_g\left(\frac{n+1}{n}c^{2n+1} + d^{2n+1}\right)\right]$$

$$\left.+ \epsilon_g P_n\left[\epsilon_0\left(c^{2n+1} + \frac{n+1}{n}d^{2n+1}\right) - \frac{n+1}{n}\epsilon_g(d^{2n+1} - c^{2n+1})\right]\right\}$$

$$Q_n = \frac{n+1}{n}\epsilon_p\left[\epsilon_p(a^{2n+1} - b^{2n+1})(c^{2n+1} - b^{2n+1})\right.$$

$$\left.- \epsilon_0\left(c^{2n+1} + \frac{n+1}{n}b^{2n+1}\right)\left(\frac{n}{n+1}b^{2n+1} + a^{2n+1}\right)\right]$$

Enhanced Radiation from an Antenna

$$P_n = \frac{n+1}{n}\left[\epsilon_p(a^{2n+1} - b^{2n+1})\left(b^{2n+1} + \frac{n+1}{n}c^{2n+1}\right)\right.$$
$$\left. + \epsilon_0(b^{2n+1} - c^{2n+1})\left(b^{2n+1} + \frac{n+1}{n}a^{2n+1}\right)\right]$$

$$\rho_n = \frac{k_0^{2n+1}}{(2n+1)[1 \cdot 3 \cdot 5 \cdots (2n-1)]^2}$$

Direct inspection of formula (6-39) shows two features of the radiated field. F_n is proportional to ϵ_p and in a collision free plasma ($\nu/\omega = 0$; ν: collision frequency), we therefore have $F_n = 0$ for $\omega = \omega_e$. This antenna exhibits an antiresonance at the plasma frequency. When the collisions are small ($\nu/\omega \ll 1$), the radiated field is very small for $\omega = \omega_e$ and this results in a blackout at the plasma frequency which corresponds, in certain conditions, to the well known blackout phenomenon observed with satellites. The second feature is that the real part of Δ_n is quadratic in $\epsilon_p = \epsilon_0(1 - \omega_e^2/\omega^2)$ and that it possesses two zeros for $\omega < \omega_e$. Each multipolar spherical wave of order n exhibits thus two resonances corresponding to these zeros. The theoretical radiated field computed in a situation corresponding to experiments is shown in figure 6.13.

The active radiated power P_a has the following form:

$$P_a = \pi \lim_{r \to \infty} r^2 \int_0^\pi E_\theta H_\phi^* \sin\theta \, d\theta = \pi \sqrt{\frac{\epsilon_0}{\mu_0}} \sum_{n=1,3,\ldots} \frac{2n(n+1)}{2n+1} |F_n|^2 \quad (6\text{-}40)$$

This power is therefore zero at the anti-resonance in a collisionless plasma and goes through a maximum whenever $|F_n|$ goes through a maximum, i.e. for each of the multipolar resonances. The radiation conductance C is immediately deduced from P_a:

$$C = \frac{P_a}{2\phi_0^2} = \frac{\pi}{2\phi_0^2}\sqrt{\frac{\epsilon_0}{\mu_0}} \sum_{n=1,3,\ldots} \frac{2n(n+1)}{2n+1}|F_n|^2 \quad (6\text{-}41)$$

Problem: Consider the simple case for which

$$\left(\frac{b}{c}\right)^2 \ll 1 \quad \text{and} \quad \left(\frac{s}{a}\right)^2 = \left(\frac{b-a}{a}\right)^2 \ll 1$$

and show that, if the influence of the glass wall is neglected ($\epsilon_g = \epsilon_0$), the resonances—i.e. the zeros of Δ_n—are approximately given by

$$\left(\frac{\omega_e}{\omega}\right)^2_{Rn} \simeq 1 + \frac{n+1}{n}\left(1 + \frac{(n+1)^2}{n}\frac{s}{a}\right) \quad (6\text{-}42)$$

and

$$\left(\frac{\omega_e}{\omega}\right)^2_{Rn'} \simeq 1 + \frac{n+1}{n} + \frac{n}{n+1} + \frac{a}{(n+1)s} \quad (6\text{-}43)$$

The first set of resonances, the unprimed set, is only slightly influenced by the sheath thickness s and is the only one which exists if there is no sheath. This type of resonance is due to the exchange of energy between the outer vacuum (capacitive medium characterized by ϵ_0) and the plasma layer (inductive medium characterized by $\epsilon_p < 0$). We note that when $s = 0$, the resonances (6-42) are precisely those of a plasma sphere placed in vacuum in the quasistatic approximation.

The second set of resonances, the primed set, is very strongly influenced by the sheath thickness s. This type of resonance is mostly due to an exchange of energy between the plasma layer and the sheath. These resonances can become very important. A dielectric layer placed around the antenna has, naturally, the same influence as the sheath.

Influence of non-zero temperature

As already underlined when examining the metallic probe, the developments of chapter 5 amply show that if we wish to calculate the secondary or temperature spectrum in a realistic fashion, we must not only consider a hot plasma but also the exact density profile. This gives rise to a complicated numerical calculation and the observations with the antenna indicate that such a theoretical complexity is not needed for the interpretation of the experimental results.

b. Experiments

The $|E_\theta|$ experimentally observed at $\theta = 90°$ and $r = 10$ cm is shown in figure 6.12 as a function of the discharge current I_d of the tube. This I_d is proportional to the average electron density $\langle N \rangle$, and is thus proportional to $\Omega^2 = \omega_e^2/\omega^2$ since the experiments are carried out at a given operating frequency $\omega/2\pi = 300$ MHz. Each oscillogram of figure 6.12 is characterized by a given value of the DC bias potential V_0 and thus by a given value of the sheath thickness s. The value of s is estimated through

$$s = Cr_D \left(\frac{|eV_0|}{KT_e}\right)^{\frac{3}{4}} \tag{6-44}$$

where r_D is the Debye length, T_e the electron temperature and C a constant which is taken approximately equal to 2[204]. Note that the field which is radiated by the antenna in the absence of plasma ($\Omega^2 = 0$) is very small, since the radius a of the antenna is much smaller than the vacuum wavelength λ_0.

The theoretical $|E_\theta|$ is also computed for $\theta = 90°$, $r = 10$ cm in order to compare it to observational data. Figure 6.13 shows these theoretical results for 3 typical values of the sheath thickness ($s = 5$, 10 and $15r_D$). The different $n = 1, n = 3$ etc. peaks are due to the resonance phenomenon between the plasma and the outer vacuum while the different $n = 1'$, $n = 3'$ etc. peaks are the corresponding resonances due to the exchange of energy involving the plasma and the plasma sheath. These last resonances

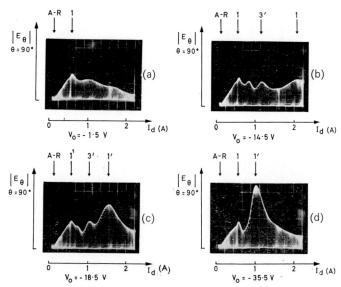

Figure 6.12 Oscillograms giving $|E_\theta|$ at $\theta = 90°$, $r = 10$ cm as a function of $I_d \propto \langle N \rangle \propto \Omega^2 = \omega_e^2/\omega^2$. The vertical scale is linear and each curve is characterized by the dc potential V_o of the antenna with respect to the plasma

are strongly influenced by the sheath thickness as predicted by the simplified relation (6-43) and we remark that they are considerably shifted towards lower values of Ω^2 when s increases. This is entirely confirmed by the experimental data of figure 6.12 where resonance $1'$ first appears on figure 6.12b but is already dominant on figure 6.12c.

The anti-resonance is observed on figure 6.12 at $\omega = \omega_e$ as predicted theoretically and the value of ω_e is checked by an independent method.

The field radiated by the antenna in the absence of plasma ($\Omega^2 = 0$) is very small when compared to the $n = 1'$ resonance field due to the plasma and its sheath (figure 6.12d). For this value of the plasma density, the radiated field is 20 times stronger than the field in the absence of plasma, and the radiated power is thus 400 times greater than without plasma. This is owing to the fact that, at resonance, the reactance presented by the antenna-plasma system is practically zero and the impedance is thus

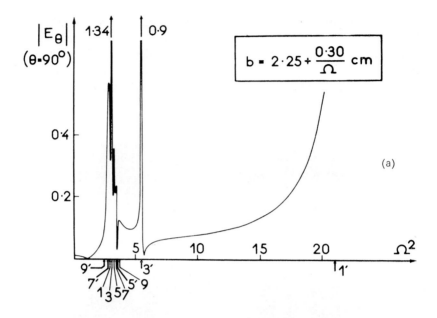

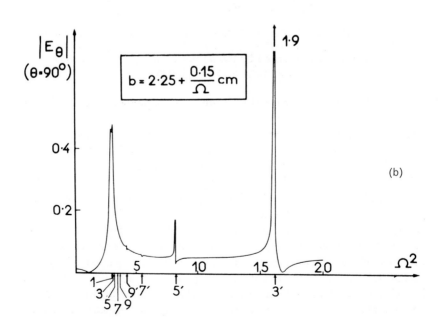

Enhanced Radiation from an Antenna

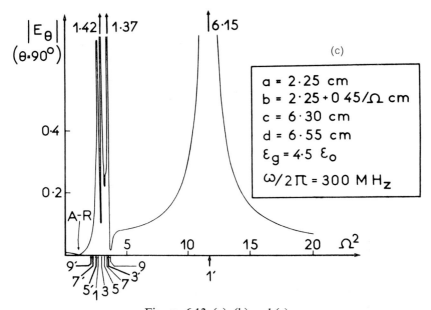

Figure 6.13 (a), (b) and (c)
Figure 6-13. Theoretical curves of the radiated $|E_\theta|$ ($\theta = 90°$, $r = 10$ cm, $\phi_0 = 1$ V) versus $\Omega^2 = \omega_e^2/\omega^2$. The curves are characterized by increasing sheath thickness.

resistive (in this case the radiation resistance is very small and approximately equal to 0·3 Ω). It is therefore clear that, for this resonance and for a given voltage of the generator, the radiated power is limited by the internal impedance of the generator (50 Ω in the case of the experiment). The corresponding theoretical curve (figure 6.13c) must necessarily show a radiated field at resonance which is relatively much stronger than the experimental one since the theory was performed assuming a constant difference of potential $2\phi_0$ between the lips of the hemispheres and that this corresponds to the case of a generator with zero internal impedance. Equation (6-41) clearly shows that the radiation resistance is inversely proportional to $|F_n|^2$ and that it is thus considerably lowered at resonance.

The curves of figure 6.12 are obtained at rather low incident power ($\simeq 0·1 W$). Experiments performed at higher power ($\simeq 10 W$) show that the plasma can then be maintained by RF ionization in a resonant state without the auxiliary DC discharge creating the plasma (section 6.3c). This behaviour is closely connected to that analysed in the experiment of section 2.3.

This plasma coated-antenna being reciprocal, it can be used as a frequency-sensitive receiving antenna. The phenomena observed when V_0 is varied are identical to those observed in emission. We should expect that in reception the performances be limited by the plasma noise but with the receiver used (having a noise level of 18 dB) there was no systematic difference in the noise observed with or without plasma around the antenna.

Finally, it should be stressed that this model is immediately applicable to a satellite immersed in a plasma or surrounded by a re-entry plasma layer if ν/ω is sufficiently low.

Influence of a steady magnetic field

When a steady magnetic field is applied to this antenna, one must expect to encounter difficulties similar to those described in the case of the metallic resonance probe. The effect of a steady magnetic induction $\mathbf{B}_0$ oriented along the anode-cathode axis of the tube has been investigated. For low values of $\mathbf{B}_0$ ($<2 \times 10^{-4}$ T) there is practically no influence on the phenomena observed. When $\mathbf{B}_0$ increases, the resonances are damped out and considerably deformed. It is not possible to conclusively observe the splitting of the resonances which is expected on the basis of the anisotropy of the plasma (section 4.2b). The complexity of the phenomena observed is due to the non-uniformity and asymmetry of the density distribution around the antenna when $\mathbf{B}_0 \neq 0$. When $\mathbf{B}_0$ is only 20×10^{-4} T, the discharge concentrates itself more and more along the anode-cathode axis and becomes unstable because the antenna is an obstacle which the plasma surrounds in an erratic fashion. Again, this is a clear indication that a worth-while theoretical treatment of the influence of $\mathbf{B}_0$ on plasma coated antennas or metallic resonance probes must take the influence of $\mathbf{B}_0$ on the sheath into account.

c. Non-linear behaviour

When sufficient HF power is fed to the plasma-coated antenna, the plasma surrounding the antenna becomes resonantly sustained, i.e. a plasma is obtained in the absence of a DC discharge, and the state of the plasma is such that the radiated field is enhanced. The emission of harmonics at 2ω and 3ω can also be observed. To explain this latter feature, the different linear resonance patterns at ω, 2ω and 3ω must be considered and when a resonance state at ω (given $\langle N \rangle$ and s) corresponds also to a resonance state at 2ω or 3ω, there is a great enhancement of the harmonic signal. The interested reader is referred to the original paper[213] for further details.

It is becoming very clear that this resonantly high frequency sustained behaviour is a general feature of plasma systems (see also sections 2.3 and 5.4d).

6.4 General considerations on resonances and anti-resonances of plasma systems

We conclude this chapter by examining some general features of the resonant behaviour of cold collisionless plasma systems in the quasi-static approximation.

We have seen that the plasma slab-condenser system (chapter 2) displays a resonance ($Z = V/J = 0$) for $\omega = \omega_e \theta$ ($0 < \theta < 1$) and an anti-resonance ($Z = \infty$) for $\omega = \omega_e$. The analysis of the metallic resonance probe in different geometries (section 6.1) gives completely analogous results: a resonance for $\omega = \omega_e \theta$ ($0 < \theta < 1$) and an anti-resonance for $\omega = \omega_e$.

In the scattering of an electromagnetic wave by a homogeneous plasma cylinder in vacuum, there is resonance scattering for $\omega = \omega_e/\sqrt{2}$ while a column of vacuum in an infinite plasma possesses the same eigenfrequency. If there is an additional dielectric medium around the plasma column, the resonance frequency is not only altered but the system also becomes transparent (no scattered field) for another characteristic $\omega < \omega_e$ (chapter 3). It is also noted that the dielectric two-wire probe surrounded only by vacuum and plasma has a resonance for $\omega = \omega_e \delta$ ($0 < \delta < 1$) where δ depends on the geometrical characteristics of the probe and an anti-resonance for $\omega = \omega_e/\sqrt{2}$.

The question that arises is the following: Why is the first class of resonators characterized by an anti-resonance for $\omega = \omega_e$ and a resonance for $\omega_{\text{res}} = \omega_e \theta$, while the second class of resonators is characterized by an anti-resonance for $\omega \neq \omega_e$ and a resonance for $\omega_{\text{res}} = \omega_e \delta$? Also, why does the resonance scattering for $\omega = \omega_e/\sqrt{2}$ correspond to the anti-resonance of the dielectric two-wire probe in indentical geometry?

Plasma systems characterized by one coordinate

We consider the systems for which the equipotential surfaces depend only on one coordinate and where the interfaces between different media coincide with these equipotential surfaces. This question is better investigated in a special coordinate system, e.g. spherical coordinates, but we shall see that the features found are general.

The spherical metallic probe is sketched in figure 6.2. It is characterized by the fact that the equipotential surfaces depend only on r. We briefly recall some main points of the analysis of section 6.1. The solutions of the Laplace equation for $m = n = 0$ (m and n giving the angular dependence in ϕ and θ) in the two media are

$$\phi_1 = A + B/r; \qquad \phi_2 = C/r \qquad (6\text{-}45)$$

Using the continuity of the potential and of the normal component of the electric displacement at each spherical interface, the current density on the probe $J_{\mathrm{HF}} = \epsilon_0 (\partial E/\partial t)_{r=R}$ is obtained. This current displays a resonance for $\omega = \omega_e \sqrt{s/R+s}$ and an anti-resonance for $\omega = \omega_e$. Now the reason for obtaining the anti-resonance at precisely $\omega = \omega_e$ is that, when $\epsilon_p = 0$, the theorem of Gauss leads to $E = 0$ outside the plasma.

Indeed, figure 6.14 shows that, since there are no E_θ or E_ϕ components, the theorem of Gauss gives

$$\epsilon_p E_{r=R_2} R_2^2 = \epsilon_1 E_{r=R_1} R_1^2 \qquad (6\text{-}46)$$

where $r = R_1$ ($R < R_1 < R+s$) is an arbitrary spherical surface in medium 1 (dielectric or vacuum) and $r = R_2$ is an arbitrary spherical surface in the plasma. When $\epsilon_p = 0$, this entails that $E = 0$ in medium 1 outside the plasma and thus $J_{\mathrm{HF}} = 0$. This result is general in any

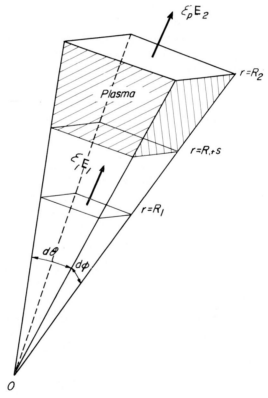

Figure 6.14 Application of Gauss's theorem to a volume element contained between θ and $\theta + d\theta$, ϕ and $\phi + d\phi$, $r = R_1$ and $r = R_2$

orthogonal curvilinear system q^1, q^2, q^3 where the potential is only a function of one coordinate q^i.

The resonance corresponds to $E_{r=R} = \infty$, i.e. infinite potentials everywhere in the system, and the geometry plays its role through R and $R + s$ because

$$\int_R^\infty E_r \, dr = 0 \qquad (6\text{-}47)$$

It is that value of ω for which (6-47) is nontrivially satisfied which corresponds to the resonance, this is only why the resonance necessarily occurs for $\epsilon_p < 0$, i.e. $\omega_{\text{res}} = \omega_e \theta$ $(0 < \theta < 1)$. Indeed, because of the theorem of Gauss, the sign of the electric displacement D_r is constant and there must exist a portion of space where $\epsilon_p < 0$ in order that E_r may change its sign. The physical meaning of equation (6-47) is that a finite applied voltage V is not balanced by a difference of potential between the probe and infinity, hence the probe draws an infinite current. Again, the result is valid for any curvilinear system where the potential is a function of a single coordinate q^i.

Plasma systems characterized by more than one coordinate

The second class of resonators is characterized by the fact that the potential is a function of more than one coordinate. Since the electric field then has more than one component, the application of the theorem of Gauss, when $\epsilon_p = 0$, does not generally lead to $\mathbf{E} \equiv 0$ outside the plasma, and thus to an anti-resonance.

The simplest spherical dielectric probe is studied as an example and sketched in figure 6.15. The excitation in the quasistatic approximation is provided by two spherical charges $(+q, -q)$ and is, for $b/a \ll 1$, preponderantly dipolar. The 'applied' potential ϕ_a and the potentials ϕ_1 and ϕ_2 induced in medium 1 and 2 are given by

$$\phi_a = \frac{qb \cos \theta}{4\pi\epsilon_0 r^2} ; \qquad \phi_1 = Ar \cos \theta ; \qquad \phi_2 = Br^{-2} \cos \theta \qquad (6\text{-}48)$$

Following the same method as in section 6.2, we obtain the capacity C between the two charged spheres (C_a: capacity of the spheres in vacuum):

$$Z = \frac{-1}{i\omega C} = \frac{-1}{i\omega}\left[\frac{1}{C_a} + \frac{1}{2\pi\epsilon_0}\frac{b^2}{a^3}\left(\frac{\epsilon_0 - \epsilon_p}{\epsilon_0/2 + \epsilon_p}\right)\right] \qquad (6\text{-}49)$$

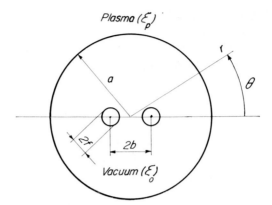

Figure 6.15 Simple spherical dielectric probe

Problem: Deduce this last expression.

The anti-resonance ($Z = \infty$) occurring for $\omega = \omega_e\sqrt{\tfrac{2}{3}}$ corresponds to infinite induced potentials everywhere and particularly to (E_i: induced electric field strength):

$$\int_{-b+f}^{b+f} E_{ri}\, dr = \infty \tag{6-50}$$

Now, for this same value of ω

$$\int_{0}^{\infty} E_{ri}\, dr = 0 \tag{6-51}$$

and $\omega = \omega_e\sqrt{\tfrac{2}{3}}$ is thus also the frequency for which the system displays a resonance on the scattering picture. A resonance is clearly determined by the fact that

$$\int_{0}^{\infty} \mathbf{E}_i \cdot d\mathbf{r}$$

is non trivially zero (this is evidently also true for equation (6-47) since $\mathbf{E} \equiv 0$ between 0 and R) and that the corresponding fields are infinite. If we now place two electrodes at a finite distance, the difference of potential is infinite since it is given by an expression of the type (6-50) and we observe an anti-resonance between these electrodes.

It directly follows that the resonance ($Z = 0$) observed between the two electrodes depends not only on the plasma and the dielectrics present but also on the geometrical characteristics of the electrodes. This is precisely the case with the simple spherical dielectric probe whose impedance Z is given by equation (6-49). The resonance occurs at a frequency for which the plasma and the dielectric form an equivalent condenser of capacity C_{eq} balancing the capacity of the two metallic spheres, i.e. we have $C_{eq} = -C_a$. This value is not inherent in the geometry of vacuum sphere and plasma but also depends on the dimensions of the two-sphere probe. For this value we have

$$\int_{-b+f}^{b-f} E_r \, dr = 0 \tag{6-52}$$

where E_r is the total field ($E_{ra} + E_{ri}$).

Problem: Using the boundary conditions, div ($\phi\epsilon$ grad ϕ) = ϕ div (ϵ grad ϕ) + ϵ (grad ϕ)² and Ostrogradski's theorem, show that the resonance condition

$$\int_0^\infty \mathbf{E_i} \cdot \mathbf{dr} = 0 \tag{6-53}$$

is equivalent to

$$\int_\tau \epsilon E_i^2 \, d\tau = 0 \tag{6-54}$$

where τ is the total volume of plasma and dielectrics and where E_i is not trivially zero everywhere (see equation (1-48)).

Problem: Consider, in the quasi-static approximation and with a uniform applied electric field, a cylinder of plasma of radius a placed in vacuum embedded in an infinite plasma and consider also a cylinder of vacuum of radius a. Show that when $\epsilon_p = 0$, $\mathbf{E} \neq 0$ for the former system except in the vacuum at $r = a$ while $\mathbf{E} = 0$ in the whole vacuum of the latter system.

APPENDIX 1

Volume Theorems Useful in Stating Mathematical Boundary Conditions

It is recalled in section 1.9 that the very form of the equations describing a hot non-uniform plasma impose mathematical conditions on the physical variables at the boundary between a plasma and another medium or at the surface of discontinuity between two different idealized plasmas.

We shall there state the form which some useful volume theorems take at such surfaces of discontinuity. Using these formula, the reader can readily establish the mathematical boundary condition pertaining to many an equation.

1. Gradient theorem

If V is a volume limited by a closed surface S and if the unit vector normal in each point to S and directed towards the *outside* is noted $\mathbf{u}_n$, the gradient theorem is then expressed by

$$\iiint_V \operatorname{grad} \phi \, dV = \oiint_S \phi \mathbf{u}_n \, dS \qquad (A\text{-}1)$$

Consider now a surface of discontinuity between two media 1 and 2 (figure A-1) and a cylinder having bases of area dS in 1 and 2 respectively and having a height Δl along the normal to the boundary surface. Applying (A-1) to this cylinder of volume $dS\Delta l$ and letting Δl tend to zero, we obtain

$$\lim_{\Delta l \to 0} \Delta l \operatorname{grad}_n \phi = \phi_2 - \phi_1 \qquad (A\text{-}2)$$

where $\operatorname{grad}_n \phi$ is the component of $\operatorname{grad} \phi$ along the unit vector $\mathbf{u}_{n2}$ normal to the boundary surface ($\operatorname{grad}_n \phi = \operatorname{grad} \phi \cdot \mathbf{u}_{n2}$) and where ϕ_1 and ϕ_2 are respectively the values of ϕ on side 1 and side 2 of this surface.

Formula (A-2) shows the role played by a non-zero value of $\lim_{\Delta l \to 0} \Delta l \operatorname{grad}_n \phi$ and how an equation containing a gradient must be integrated across a boundary surface limiting two different media.

Appendix

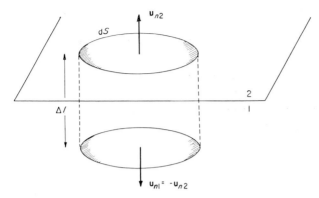

Figure A.1 Boundary surface between media 1 and 2 with an elementary cylinder having base dS and height Δl

2. Divergence theorem

With the same notations as for the previous case, we have the following theorem for the divergence of a vector **A** (theorem of Gauss-Ostrogradski):

$$\iiint_V \operatorname{div} \mathbf{A}\, dV = \oiint_S \mathbf{A} \cdot \mathbf{u}_n\, dS \tag{A-3}$$

Applying (A-3) to the cylinder defined in figure A.1, we obtain the corresponding surface expression of the theorem:

$$\lim_{\Delta l \to 0} \Delta l \operatorname{div} \mathbf{A} = A_{n2} - A_{n1} \tag{A-4}$$

where A_{n1} and A_{n2} are respectively the normal component of **A** as measured along $\mathbf{u}_{n2}$ on side 1 and side 2 of the boundary surface ($A_n = \mathbf{A} \cdot \mathbf{u}_{n2}$).

Formula (A-4) shows how an equation containing a divergence must be integrated across a surface of discontinuity.

3. Curl theorem

Using again the notations previously defined, we have the following theorem for the curl of a vector **A**

$$\iiint_V \operatorname{curl} \mathbf{A}\, dV = -\oiint_S \mathbf{A} \times \mathbf{u}_n\, dS \tag{A-5}$$

Consider again an elementary cylinder of height Δl at the boundary between media 1 and 2 and $\mathbf{A} = A_n + A_x + A_y$ as defined in figure A.2

Applying (A-5) to this cylinder and letting Δl tend to zero, we obtain

$$\lim_{\Delta l \to 0} \Delta l \, \text{curl}_x \mathbf{A} = -(A_{y2} - A_{y1}) \tag{A-6}$$

$$\lim_{\Delta l \to 0} \Delta l \, \text{curl}_y \mathbf{A} = (A_{x2} - A_{x1}) \tag{A-7}$$

Formulas (A-6) and (A-7) indicate how an equation containing a curl must be integrated across a boundary surface.

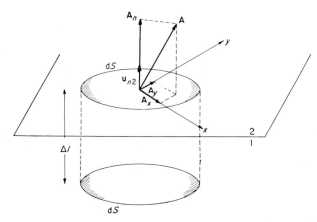

Figure A.2 Definition of $\mathbf{A} = \mathbf{A}_n + \mathbf{A}_x + \mathbf{A}_y$ on the boundary between media 1 and 2

4. Green's theorem

Considering two arbitrary scalar functions ϕ and ψ and using the previous notations, Green's theorem is

$$\iiint_V (\phi \nabla^2 \psi - \psi \nabla^2 \phi) \, dV = \oiint_S \left(\phi \frac{\partial \psi}{\partial n} - \psi \frac{\partial \phi}{\partial n} \right) dS \tag{A-8}$$

with $\partial \psi / \partial n = \text{grad } \psi \cdot \mathbf{u}_n$.

This theorem applied to the cylinder of figure A.1 leads to the following boundary condition

$$\lim_{\Delta l \to 0} \Delta l (\phi \nabla^2 \psi - \psi \nabla^2 \phi) = \left(\phi \frac{\partial \psi}{\partial n} - \psi \frac{\partial \phi}{\partial n} \right)_2 - \left(\phi \frac{\partial \psi}{\partial n} - \psi \frac{\partial \phi}{\partial n} \right)_1 \tag{A-9}$$

with $\partial \psi / \partial n = \text{grad } \psi \cdot \mathbf{u}_{n2}$.

Appendix

Problem. Consider two hot uniform plasmas of temperature T and of respective equilibrium densities $\langle N_1 \rangle$ and $\langle N_2 \rangle$ which are subject to a high frequency perturbation. Show that at the boundary between these two plasmas described by equations (5-2), (5-3) and (5-4), we have the following boundary conditions if we assume that there is no high frequency surface charge:

$$E_{n2} = E_{n1} \tag{A-10}$$

$$\langle N_2 \rangle \mathbf{u}_{n2} = \langle N_1 \rangle \mathbf{u}_{n1} \tag{A-11}$$

$$\frac{\langle N_2 \rangle^{1/\gamma}}{n_2} = \frac{\langle N_1 \rangle^{1/\gamma}}{n_1} \tag{A-12}$$

References

Books

1. W. P. Allis, S. J. Buchsbaum and A. Bers, *Waves in Anisotropic Plasmas*, M.I.T. Press, Cambridge, Mass., 1963.
2. L. Spitzer, Jr., *Physics of Fully Ionized Gases*, 2nd ed., Interscience, New York, 1962.
3. T. H. Stix, *The Theory of Plasma Waves*, McGraw-Hill, New York, 1962.
4. J. F. Denisse and J. L. Delcroix, *Plasma Waves*, Interscience, London, 1963; *Théorie des Ondes dans les Plasmas*, Dunod, Paris, 1959.
5. V. L. Ginzburg, *Propagation of Electromagnetic Waves in Plasma*, North-Holland, Amsterdam, 1961.
6. W. B. Thompson, *An Introduction to Plasma Physics*, Pergamon, Oxford, 1962.
7. B. D. Fried and S. D. Conte, *The Plasma Dispersion Function*, Academic Press, New York, 1961.
8. F. Llewellyn-Jones, *Ionisation and Breakdown in Gases*, Methuen, London, 1957.
9. P. Morse and H. Feshbach, *Methods of Theoretical Physics*, McGraw-Hill, New York, 1953.
10. D. C. Montgomery and D. A. Tidman, *Plasma Kinetic Theory*, McGraw-Hill, New York, 1964.
11. A. Sommerfeld, *Electrodynamics*, Vol. III of Lectures on Theoretical Physics, Academic Press, New York, 1954.
12. J. L. Delcroix, *Physique des Plasmas*, Vol. I, Dunod, Paris, 1963; Vol. II, 1966; *Plasma Physics*, Wiley, London, 1965.
13. J. E. Drummond (Ed.), *Plasma Physics*, McGraw-Hill, New York, 1961.
14. H. Grad, 'Principles of the Kinetic Theory of Gases' in *Handbuch der Physik* (Ed. S. Flügge), Springer-Verlag, 1958, pp. 205–294, Vol. 12.
15. J. C. Slater, *Microwave Electronics*, Van Nostrand, New York, 1950.
16. M. A. Heald and C. B. Wharton, *Plasma Diagnostics with Microwaves*, Wiley, New York, 1965.
17. C. H. Papas, *Theory of Electromagnetic Wave Propagation*, McGraw-Hill, New York, 1965.
18. P. Poincelot, *Précis d'électromagnétisme théorique*, Dunod, Paris, 1963.
19. G. Bekefi, *Radiation Processes in Plasmas*, Wiley, New York, 1966.

Papers and reports

20. A. G. Sitenko and K. N. Stepanov, *Soviet Phys. J.E.T.P.*, **31**, 642 (1956).
21. P. E. Vandenplas, *Oscillations de plasmas finis, inhomogènes et de température non nulle*; Dissertation de Doctorat, Université de Bruxelles, April 1961.

22. P. E. Vandenplas and R. W. Gould, *Proc. Vth Intern. Conf. on Ionization Phenomena in Gases*, Munich, 1961, **II**, 1470, North-Holland, Amsterdam, 1962.
23. P. E. Vandenplas and R. W. Gould, *Physica*, **28**, 357 (1962).
24. L. Oster, *Rev. Mod. Phys.* **32**, 141 (1960).
25. E. Åström, *Arkiv Fysik*, **19**, 163 (1961).
26. P. Weissglas, *Plasma Phys. (J. Nucl. Energy*, Part C), **4**, 329 (1962).
27. A. M. Messiaen and P. E. Vandenplas, *Physica*, **30**, 303 (1964).
28. J. Taillet, *Compt. Rend. VI Conf. Intern. Phénomènes d'Ionisation dans les Gaz*, Paris (1963).
29. J. Taillet, *Ph.D. Thesis*, University of Paris, Juin 1964 (Published as Rappt. CEA-R2502, C.E.N. Saclay, 1964); 3.1.7(7), *Proc. VIIth Intern. Conf. Ionization Phenomena in Gases*, Belgrade, August 1965.
30. A. Brunet and J. Taillet, *J. Phys.* (Paris) **26**, 520 (1965).
31. R. W. Gould, *Bull. Am. Phys. Soc.*, **8** (Series II), 170 (1963).
32. F. C. Shure, *Plasma Phys. (J. Nucl. Energy*, Part C) **6**, 1 (1964).
33. R. B. Hall, *Am. J. Phys.*, **31**, 696 (1963).
34. A. M. Messiaen and P. E. Vandenplas, *Plasma Phys. (J. Nucl. Energy*, Part C) **4**, 267 (1962).
35. P. E. Vandenplas and A. M. Messiaen, *Plasma Phys. (J. Nucl. Energy* Part C) **6**, 71 (1964).
36. A. M. Messiaen, '*Comparaison entre l'étude expérimentale de plasmas finis cylindriques et les résultats théoriques dans l'approximation de la permittivité équivalente*', Dissertation de Doctorat, Université de Bruxelles, Mai 1963.
37. T. R. Kaiser and R. L. Closs, *Phil. Mag.*, **43**, 1 (1952).
38. F. W. Crawford, *Plasma Phys. (J. Nucl. Energy*, Part C) **5**, 69 (1963).
39. F. W. Crawford, G. S. Kino, S. A. Self and J. Spalter, *J. Appl. Phys.*, **34** 2186 (1963).
40. W. M. Leavens, *Resonant Scattering of Microwaves*, Convair Report, California (1960).
41. H. Shapiro, *Antenna Laboratory Tech. Rept.*, No. 11, California Institute of Technology (1957).
42. G. D. Boyd, L. M. Field and R. W. Gould, *Phys. Rev. (Letter)*, **109**, 1393 (1958).
43. A. Dattner, *Ericsson Technics*, **2**, 309 (1957).
44. A. M. Messiaen and P. E. Vandenplas, *Physica*, **28**, 537 (1962).
45. A. M. Messiaen and P. E. Vandenplas, *Phys. Letters*, **2**, 193 (1962).
46. P. E. Vandenplas and A. M. Messiaen, *Nucl. Fusion*, **5**, 47 (1965).
47. A. M. Messiaen and P. E. Vandenplas, *Nucl. Fusion*, **5**, 56 (1965).
48. P. M. Platzman and H. T. Ozaki, *J. Appl. Phys.*, **31**, 1597 (1960).
49. P. S. Epstein, *Rev. Mod. Phys.*, **28**, 3 (1956).
50. A. M. Messiaen and P. E. Vandenplas, *Phys. Rev.*, **149**, 131 (1966).
51. R. J. Bickerton and A. von Engel, *Proc. Phys. Soc.*, **69B**, 468 (1956).
52. F. W. Crawford, G. S. Kino and A. B. Cannara, *J. Appl. Phys.*, **34**, 3168 (1963).
53. L. D. Landau, *J. Phys. (U.S.S.R.)*, **10**, 25 (1946).
54. T. Wasserab, *Z. Naturforsch.*, **2a**, 575 (1947).
55. L. Beckmann, *Proc. Phys. Soc.*, **61**, 515 (1948).
56. W. C. Roentgen, *Ann. Phys.*, **35**, 264 (1888) and **40**, 93 (1890).
57. A. Eichenwald, *Ann. Phys.*, **11**, 421 (1903).

58. P. Davidson, *Proc. Phys. Soc.*, **B67,** 159 (1954).
59. G. Bryant and N. Franklin, *Proc. Phys. Soc.*, **81,** 531 and 790 (1963).
60. P. E. Vandenplas and R. W. Gould, "Study of electromagnetic interactions in plasmas", *Contract U.S. Army Signal Res. Develop. Lab. Quart. Progr. Report No.* **3,** California Institute of Technology, 1960.
61. P. E. Vandenplas and R. W. Gould, *Nucl. Fusion*, 1962 Supplement Part III, 1155; *Plasma Phys. (J. Nucl. Energy*, Part C) **6,** 449 (1964).
62. P. E. Vandenplas and A. M. Messiaen, *Plasma Phys. (J. Nucl. Energy*, Part C) **6,** 459 (1964).
63. L. Tonks, *Phys. Rev.*, **37,** 1458 (1931); *Phys. Rev.*, **38,** 1219 (1931).
64. N. Herlofson, *Arkiv Fysik*, **3,** 247 (1951).
65. R. N. Franklin, *J. Elec. Controls*, **17,** 513 (1964).
66. M. Feix, *Phys. Letters*, **12,** 316 (1964).
67. R. W. Gould, *Quart. Status Rept.* No. 16 *Contract, Office Naval Res.*, Electron Tube and Microwave Laboratory, California Institute of Technology, (1957).
68. E. M. Barston, *Microwave Laboratory Rept.* No. 1064, Stanford University (1963).
69. E. M. Barston, *Phys. Fluids*, **6,** 828 (1963).
70. E. M. Barston, *Ann. Phys. (New York)*, **29,** 282 (1964).
71. P. A. Sturrock, *Ann. Phys. (New York)*, **4,** 306 (1958).
72. F. E. Low, *Proc. Roy. Soc. (London)*, **A248,** 283 (1958).
73. F. W. Crawford and G. S. Kino, *Compt. Rend. Acad. Sci. (Paris)*, **256,** 1939 and 2798 (1963).
74. C. R. Obermann, *Bull. Am. Phys. Soc.*, Series 2, **5,** 364 (1960).
75. C. R. Obermann, *Matt*-57, *Proj. Matterhorn*, Princeton University, 1960.
76. N. G. Van Kampen, *Physica*, **23,** 641 (1957).
77. J. C. Nihoul and P. E. Vandenplas, *Plasma Phys. (J. Nucl. Energy*, Part C), **7,** 341 (1965).
78. R. N. Franklin and G. H. Bryant, *Proc. Inst. Elec. Engrs.*, **110,** 1709 (1963).
79. J. V. Parker, J. C. Nickel and R. W. Gould, *Phys. Fluids*, **7,** 1489 (1964).
80. J. C. Nickel, J. V. Parker and R. W. Gould, *Phys. Rev. Letters*, **11,** 183 (1963).
81. J. C. Nickel, '*Experimental Study of Plasma Wave Resonances in a Hot Nonuniform Plasma Column*', Ph.D. Dissertation, California Institute of Technology, May 1964.
82. J. V. Parker, '*Theory of Plasma Wave Resonances in a Hot Nonuniform Plasma*', Ph.D. Dissertation, California Institute of Technology, June 1964.
83. S. C. Brown *et al.*, *Tech. Rept.* No. 66, *Res Lab Electron.*, Massachusetts Institute of Technology (1948).
84. S. C. Brown and D. J. Rose, *J. Appl. Phys.*, **23,** 711, 719 and 1028 (1952).
85. S. J. Buchsbaum and S. C. Brown, *Phys. Rev.*, **106,** 196 (1957).
86. S. C. Brown, *Proc. Second Intern. Conf. Peaceful Uses of Atomic Energy*, (Geneva), **32,** 394 (1958).
87. S. J. Buchsbaum, L. Mower and S. C. Brown, *Phys. Fluids*, **3,** 806 (1960).
88. B. Agdur and B. Enander, *J. Appl. Phys.*, **33,** 575 (1962).
89. K. I. Thomassen, *J. Appl. Phys.*, **34,** 1622 (1963); **36,** 3642 (1965).
90. D. Bohm and E. P. Gross, *Phys. Rev.*, **75,** 1851 and 1864 (1949).
91. D. Bohm and E. P. Gross, *Phys. Rev.*, **79,** 992 (1950).
92. D. Romell, *Nature*, **167,** 243 (1951).

93. R. E. B. Mackinson and D. M. Slade, *Australian J. Phys.*, **7**, 268 (1954).
94. G. Keitel, *Proc. IRE*, **43**, 1481 (1955).
95. V. B. Gil'denburg, *J. Tech. Phys.*, (U.S.S.R), **34**, 372 (1964).
96. A. Dattner, *Proc. Vth Intern. Conf. Ionization Phenomena in Gases*, Munich 1961, Vol. **II**, 1477, North-Holland, Amsterdam, 1962.
97. A. Dattner, *Ericsson Techn.*, **8**, 1 (1963); *Phys. Rev. Letters*, **10**, 205 (1963).
98. F. I. Boley, *Nature*, London **182**, 791 (1958).
99. W. D. Herschberger, *J. Appl. Phys.*, **31**, 417 (1960).
100. F. I. Boley, *J. Appl. Phys.*, **31**, 1692 (1960).
101. J. Willis and I. Petroff, *IRE Trans.* PGMTT **10**, 395 (1962).
102. A. M. Messiaen and P. E. Vandenplas, *Physica*, **30**, 2309 (1964).
103. F. W. Crawford, *J. Appl. Phys.*, **35**, 1365 (1964).
104. G. H. Bryant and R. N. Franklin, *Proc. Phys. Soc.*, **83**, 971 (1964).
105. A. M. Messiaen, *Physica*, Letter, **29**, 1117 (1963).
106. R. A. Stern, *J. Appl. Phys.*, **34**, 2562 (1963).
107. H. J. Schmitt, *Appl. Phys. Letters*, **4**, 111 (1964).
108. R. W. Gould, *Proc. Linde Conf.*, *Plasma Oscillations*, Indianapolis (1959); *Bull. Am. Phys. Soc.*, **5**, 322 (1960).
109. J. V. Parker, *Techn. Rept.* No. 20, Nonr. 220(13), May 1963, California Institute of Technology.
110. P. Weissglas, *Phys. Rev. Letters*, **10**, 206 (1963).
111. R. B. Hall, *Proc. Vth Intern. Conf. Ionization Phenomena in Gases*, Paris, **3**, 35 (1963).
112. F. C. Hoh, *Phys. Rev.* **133A**, 1016 (1964).
113. V. B. Gil'denburg, *Sov. Phys. JETP*, **18**, 1359 (1964).
114. P. E. Vandenplas and A. M. Messiaen, *Proc. VIth Intern. Conf. Ionization Phenomena in Gases*, (Paris), **3**, 33 (1963).
115. P. Weissglas, *Plasma Phys.* (*J. Nucl. Energy*, Part C), **6**, 251 (1964).
116. J. A. Fejer, *Phys. Fluids*, **7**, 439 (1964).
117. R. W. Gould, *Rept.* No. 1, *Contract* DA36-039 SC-85317, California Institute of Technology (1960).
118. F. W. Crawford, *Phys. Letters*, **5**, 244 (1963).
119. F. W. Crawford and G. S. Kino, *Proc. VIth Intern. Conf. Ionization Phenomena in Gases* (Paris), **3**, 51 (1963).
120. T. G. Killian, *Phys. Rev.*, **35**, 1238 (1930).
121. P. E. Vandenplas, *Rept.* No. 6, *Lab. Phys. Plasmas.* Ecole Royale Militaire, Brussels, 1963.
122. J. V. Parker, *Phys. Fluids*, **6**, 1657 (1963).
123. L. Tonks and I. Langmuir, *Phys. Rev.*, **34**, 876 (1929).
124. B. Agdur, B. Kerzar and F. Sellberg, *Phys. Rev.*, **128**, 1 (1963).
125. B. Agdur, *Rept. Air Force Contract* AF61 (052) 552, Stockholm Institute of Technology, January 31, 1964.
126. C. D. Lustig, *Phys. Letters*, **9**, 315 (1964).
127. P. Weissglas, *Arkiv Fysik*, **26**, 185 (1964).
128. R. W. Gould, *Paper* 4.5.1. (16) *Proc. 7th Intern. Conf. Phenomena Ionized Gases*, Belgrade 1965.
129. J. H. Malmberg, C. B. Wharton and W. E. Drummond, '*Plasma Physics and Controlled Nuclear Fusion Research*', 485, Vol. I, *Proc. Culham Conf.* 1965. I.A.E.A. Vienna, 1966.
130. J. C. Nihoul and P. E. Vandenplas, *Paper* 4.4.1.(18), *Proc. 7th Intern. Conf. Phenomena Ionized Gases*, Belgrade 1965.

131. S. J. Buchsbaum and A. Hasegawa, *Phys. Rev. Letters*, **12**, 685 (1964).
132. S. J. Buchsbaum and A. Hasegawa, *Phys. Rev.*, **143**, A 303 (1966).
133. J. C. Nihoul, *Plasma Phys.*, **9**, 351 (1967).
134. B. O'Brien, R. W. Gould and J. Parker, *Phys. Rev. Letters*, **14**, 630 (1965).
135. B. O'Brien, *Plasma Phys.*, **9**, 369 (1967).
136. A. W. Trivelpiece and R. W. Gould, *J. Appl. Phys.* **30**, 1784 (1959).
137. G. S. Kino et al., *Proc. VIIth Intern. Conf. Phenomena in Ionized Gases* (Section 4.4.1.) Belgrade 1965.
138. E. A. Jackson and M. Raether, *Phys. Fluids*, **9**, 1257 (1966).
139. R. N. Franklin, *Phys. Letters*, **22**, 580 (1966).
140. R. W. Huggins and M. Raether, *Phys. Rev. Letters*, **17**, 745 (1966).
141. P. E. Vandenplas and A. M. Messiaen, *Plasma Phys..* **9**, 511 (1967).
142. E. M. Barston, *Phys. Rev.*, **139**, A394 (1965).
143. K. Mitani, H. Kubo and S. Tanaka, *J. Phys. Soc.* Japan, **19**, 211 (1964).
144. F. W. Crawford, G. S. Kino and H. H. Weiss, *Phys. Rev. Letters*, **13**, 229 (1964).
145. H. J. Schmitt, G. Meltz and P. J. Freyheit, *Phys. Rev.*, **139**, A1432 (1965).
146. R. S. Harp, *Appl. Phys. Letters*, **6**, 51 (1965).
147. G. Landauer, *Proc. 5th Intern. Conf. on Ionization Phenomena in Gases*, Munich (1961), North-Holland, Amsterdam Vol 1, 389.
148. A. Hasegawa, *Phys. Fluids*, **8**, 761 (1965).
149. S. Gruber and G. Bekefi, *Paper 4.4.1.* (16), *Proc. 7th Intern. Conf. on Phenomena in Ionized Gases*, Belgrade 1965.
150. C. Azevedo, *Ph.D. Thesis*, Dept. Phys., Massachusetts Institute of Technology (1966).
151. I. B. Berstein, *Phys. Rev.*, **109**, 10 (1958).
152. G. A. Pearson, *Phys. Fluids*, **9**, 2454 (1966).
153. M. P. H. Weenink and J. Rem, *Rijnhuizen Rept.* **67-37**, Jutphaas, 1967.
154. H. L. Frisch and G. A. Pearson, *Phys. Fluids*, **9**, 2464 (1966).
155. T. H. Stix, *Phys. Rev. Letters*, **15**, 878 (1965).
156. C. W. Horton, Jr, *Phys. Fluids*, **9**, 815 (1966).
157. I. B. Bernstein and M. P. H. Weenink, *Rijnhuizen Rept.* **66-014**, Jutphaas, 1966.
158. H. H. Kuehl, *Phys. Rev.*, **154**, 124 (1967).
159. B. P. Koronov, *Sov Phys. Tech. Phys.*, **10**, 33 (1965).
160. R. A. Stern, *Phys. Rev. Letters*, **14**, 538 (1965).
161. A. M. Messiaen and P. E. Vandenplas; Contributed papers, 407, *8th Intern. Conf. Phenomena Ionized Gases*, Vienna, (1967).
162. A. A. Ivanov and D. D. Ryutov, *Sov. Phys., J.E.T.P.*, **48**, 684 (1965).
163. R. A. Stern and N. Tzoar, *Phys. Rev. Letters*, **15**, 485 (1965).
164. E. E. Salpeter, *Phys. Rev.*, **120**, 1528 (1960).
165. E. E. Salpeter, *Phys. Rev.*, **122**, 1663 (1961).
166. J. P. Dougherty and D. T. Farley, *Proc. Roy. Soc. (London)*, **A259**, 79 (1960).
167. J. A. Fejer, *Can. J. Phys.*, **39**, 716 (1961).
168. S. Ichimaru, D. Pines and N. Rostoker, *Phys. Rev. Letters*, **8**, 231 (1962).
169. N. M. Kroll, A. Ron and N. Rostoker, *Phys. Rev. Letters*, **13**, 83 (1964).
170. H. L. Berk, *Phys. Fluids*, **7**, 917 (1964).
171. K. Takayama, H. Ikegami and S. Miyazaki, *Phys. Rev. Letters*, **5**, 238 (1960).

172. Y. Ichikawa and H. Ikegami, *Progr. Theoret. Phys.* (*Kyoto*), **28,** 315 (1962).
173. Y. Ichikawa, *Proc. VIth Intern. Conf. Ionization Phenomena in Gases, Paris,* **4,** 125 (1963).
174. H. M. Mayer, *Proc. VIth Intern. Conf. Ionization Phenomena in Gases, Paris,* **4,** 129 (1963).
175. G. Peter, G. Mueller and H. H. Rabben, *Proc. VIth Intern. Conf. Ionization Phenomena in Gases, Paris,* **4,** 147 (1963).
176. G. von Gierke, G. Mueller, G. Peter and H. H. Rabben, *Z. Naturforsch.* **19a,** 1107 (1964).
177. H. Wimmel, *Inst. Plasma Phys. Rept.*, no 6/11, Garching, Munchen (1963); *Z. Naturforsch.*, **19a,** 1099 (1964).
178. S. M. Levitskii and I. P. Shashurin, *Sov. Phys. Tech. Phys.*, **8,** 319 (1963).
179. J. Pavkovitch and G. S. Kino, *Proc. VIth Intern. Conf. Ionization Phenomena in Gases, Paris* **3,** 39 (1963).
180. R. S. Harp, *Rept.*, **1117,** Standford University (Nov. 1963).
181. R. S. Harp and F. W. Crawford, *J. Appl. Phys.*, **35,** 3436 (1964).
182. A. Messiaen, *Rappt.* **PA-IGn/RT-288,** C.E.N. Saclay (1964).
183. A. Messiaen, *Compt. Rend. Acad. Sci.* (Paris), **259,** 1710 (1964).
184. T. Dote and T. Ichimiya, *J. Appl. Phys.*, **36,** 1866 (1965).
185. G. Peter, *Inst. Plasma Phys. Rept.* **IPP2/54,** Garching, Munchen, (1966); *Z. Naturforsch.* **22a,** 872 (1967).
186. G. Peter, *Phys. Letters*, **25a,** 171 (1967).
187. D. Lepechinsky, A. Messiaen, and P. Rolland, *Rappt.* **CEA-R2945,** C.E.N. Saclay, (1966).
188. D. Lepechinsky, A. Messiaen and P. Rolland, *Plasma Phys.*, **8,** 165 (1966).
189. J. Taillet, *J. Phys.* (Paris), **26,** 457 (1965).
190. A. Boschi and F. Magistrelli, *Nuovo Cimento*, **29,** 487 (1963).
191. D. F. Smith and G. Bekefi, *Quart. Progr. Rept.* No 74, *Res. Lab. Electron.* Massachusetts Institute of Technology (1964).
192. J. A. Waletzko, *The Study of Low Density Plasmas with Radio Frequency Dipole Probes.* M.Sc. Thesis. Dept. Phys., Massachusetts Institute of Technology (1966).
193. F. F. Chen, *Plasma Phys.*, **7,** 47 (1965).
194. D. Lepechinsky, P. Rolland and J. Taillet, *Compt. Rend. Acad. Sci.*, (Paris), **261,** 663 (1965).
195. C. W. McCutchen, *Phys. Rev.*, **112,** 1848 (1958).
196. M. A. Hellberg, Culham Lab. Mem. **M49,** Culham, 6-B (1965).
197. K. G. Budden, *Can. J. Phys.*, **42,** 90 (1964).
198. A. Messiaen, *Rappt.* **PA-IGn/RT-288 bis,** C.E.N. Saclay, (1964).
199. A. M. Messiaen and P. E. Vandenplas, *J. Appl. Phys.*, **37,** 1718 (1966).
200. A. M. Messiaen and P. E. Vandenplas, Paper 5.1.2. (22), *Proc. 7th Intern. Conf. Phenomena Ionized Gases*, Belgrade (1965).
201. F. H. C. Smith and T. W. Johnston, *J. Appl. Phys.* **37,** 4997 (1966).
202. P. E. Vandenplas, A. M. Messiaen and J. J. Papier, *Paper* THUA9, *First European Conf. Controlled Fusion and Plasma Physics*, Munich (1966).
203. A. M. Messiaen and P. E. Vandenplas, *Electron. Letters*, **3,** 26 (1967).
204. A. M. Messiaen and P. E. Vandenplas, **45,** 3367 (1967).
205. B. C. Gregory and G. Mourier, *Can. J. Phys.*, **11,** 3649 (1967).
206. A. M. Messiaen and P. E. Vandenplas, *Second European Conf. Controlled Fusion and Plasma Physics*, Stockholm (1967); *Phys. Letters*, **25A,** 339 (1967).

207. H. C. Hsuan, R. C. Ajmera and K. E. Lonngren, *Appl. Phys. Letters*, **11,** 277 (1967).
208. A. Constantin, P. Leprince, A Pointu and Y. Pomeau, *J. Phys. (Paris)*, to be published (1968).
209. H. A. H. Booth, S. A. Self and R. S. R. Sherby Harvie, *J. Elec. Control*, **4,** 434 (1958).
210. A. V. Gaponov, M. A. Miller, *Sov. Phys. JETP*, **7,** 515 (1958).
211. E. S. Weibel, In *The Plasma in a Magnetic Field*. (R. K. Landshoff, Ed). p. 60. Standford University Press, California (1958).
212. P. E. Vandenplas and A. M. Messiaen, *Phys. Letters*, **26a,** 273 (1968).
213. A. M. Messiaen, P. E. Vandenplas and G. Peter, *Electron. Letters*, **4,** 29 (1968).
214. H. Derfler and T. C. Simonen, *Phys. Rev. Letters*, **17,** 172 (1966).
215. G. Van Hoven, *Phys. Rev. Letters*, **17,** 169 (1966).
216. P. E. Vandenplas, A. M. Messiaen and J. L. Montfort. To be submitted for publication (1968).
217. E. Cannobbio and R. Crocci, *Proc. 6th Intern. Conf. on Ionization Phenomena in Gases*, (Paris), **3,** 269 (1963).
218. H. Dreicer, *Phys. Fluids*, **8,** 1568 (1965).
219. A. M. Messiaen and P. E. Vandenplas, *Plasma Phys.* (1968). To be published.
220. F. W. Crawford, R. S. Harp and T. D. Mantei, *J. Appl. Phys.*, **38,** 5077 (1967).
221. P. Leprince, Rapport L.P. 63, *Lab. Phys. Plasmas*, Centre d'Orsay, Université de Paris, 1966.

Index

Numbers in *italics* indicate the most important pages on the subject mentioned.

Afterglow 127, *165*
Agdur, B. 152, 212, 213
Ajmera, R. C. 216
Allis, W. P. 1, 210
Amplitude ratios *38*
Anisotropic dielectric rod 92, 96
Anisotropy 8, 92
  effects 93, *95* ff., *154* ff.
Antenna coated with plasma (*see* plasma coated antenna)
Approximation of moments method *19* ff.
Åström, E. 55, 102, 211
Asymmetric non-uniformity *93*, *108*, *147*
Asymmetry 14
  in a hollow cylindrical plasma *80* ff.
  in a hot plasma *154* ff.
  magnetically induced *109* ff., *147*, 159
Average (quantities) *4*
Axial propagation *1*, *158* ff.
Azevedo, C. 162, 214

Barston, E. M. 13, 212, 214
Beckmann, L. 211
Bekefi, G. 175, 210, 214, 215
Berk, H. L. 214
Bernstein, I. B. 163, 165, 314
Bers, A. 1, 210
Bessel functions, asymptotic formula 73
Bickerton, R. J. 211
Blackout phenomenon 195
Bohm, D. 212
Bohm–Gross dispersion relation 19
Boley, F. I. 134, 213
Boltzmann equation *17*
Boltzmann–Vlasov equation 17, 22, 154, 174
  in any coordinate system 25
  in cylindrical coordinates *23* ff.
  plasma slab *63* ff.
Booth, H. A. H. 216
Boschi, A. 177, 215

Boundary conditions (for plasma) 10, *28* ff., *209*
  Dirichlet–Neumann 104
  mathematical *206* ff.
Boyd, G. D. 80, 211
Brown, S. C. 41, 212
Brunet, A. 211
Bryant, G. 39, 149, 212, 213
Buchsbaum, S. J. 1, 155, 161, 163, 210, 212, 213, 214
Budden, K. G. 182, 215

Cannara, A. B. 211
Canobbio, E. 161, 216
Cavity perturbation techniques *41* ff.
Characteristic frequency *46*, *201*
Chen, F. F. 179, 189, 215
Closs, R. L. 72, 130, 211
Cold plasma approximation *3* ff., 6, *201*
Cold plasma (uniform) vs hot non-uniform plasma *160*
Cold plasma cylinder *10* ff., *72*, *78*, *79*
  and steady magnetic fields *92* ff.
  hollow *70* ff.
Cold plasma slab-condenser system *45* ff.
Collisions *5*, *47*, *53* ff.
  in hot plasma slab-condenser system *53*
Collision frequency 5
Collisionless damping *69*, *150*, *152*, *179*
Collisionless plasma *17*
Conductivity tensor (*see* equivalent conductivity tensor)
Constantin, A. 216
Conte, S. D. 67, 210
Continuity equation *4*, *18*
  in phase space 24
Crawford, F. W. 16, 80, 132, 141, 174, 211, 212, 213, 214, 215, 216
Criterion for quasi-static approximation *9*

217

Crocci, R.   161, 216
Curl theorem   207
Cut-off frequency   7
Cyclotron frequency   7, 96
Cyclotron harmonics   *161, 172*

Damping   *149* ff.
  collisional   53 ff., *149*
  collisionless (Landau)   69, *150, 152*, 179
  of temperature resonances   53 ff., *149* ff.
  radiation   11, *149* ff.
Dattner, A.   92, *131*, *141*, 211, 213
Davidson, P.   211
Debye radius (or length)   12, 14, 185
Degeneracy   102
Delcroix, J. L.   210
Denisse, J. F.   210
Derfler, H.   149, 216
Dielectric resonance probe   103, *181* ff.
  and steady magnetic field   *186, 190*
Dipole mode ($n = 1$)   *71*, 93, 102
Discharge current   36, 77
Dispersion relation   *17* ff., 19, *66*
Distribution function   17
Divergence theorem   207
Dote, T.   174, 215
Dougherty, J. P.   214
Dreicer, H.   165, 216
Drift velocity   77, 109
  high frequency effect   94, *116* ff.
Drummond, J. E.   210
Drummond, W. E.   149, 213

Eichenwald, A.   128, 211
Electron mobility   118
Electron plasma frequency   6
Elliptical plasma cylinder   91
Enander, B.   212
Enhanced radiation (of plasma coated antenna)   *191* ff.
Epstein, P. S.   211
Equilibrium quantities   4
Equivalent conductivity tensor   6 118
Equivalent permittivity   *3* ff., 6, 8
  tensorial   7, 92, *95*
Euler–Lagrange equations   *15*
Experimental set-up   30 ff., 36
Experiments, in free space   *31* ff.
  in a waveguide   *39* ff.

Farley, D. T.   214
Feix, M.   212, 214
Fejer, J. A.   132, 213
Feshbach, H.   210
Field, L. M.   211
Fluid theory (*see* hydrodynamic equations)
Franklin, N.   39, 149, 212, 213, 214
Freyheit, P. J.   214
Fried, B. D.   67, 210
Frisch, H. L.   214
Frobenius method   135

Gaponov, A. V.   216
Gil'denburg, V. B.   132, 213
Ginzburg, V. L.   8, 210
Gould, R. W.   17, 23, 43, 63, 131, 132, 140, 142, 152, 158, 211, 212, 213, 214
Grad, H.   30, 210
Gradient theorem   206
Green's theorem   208
Gregory, B. C.   215
Gross, E. P.   214
Gruber, S.   214

Hall, R. B.   132, 211, 213
Hamming's method   137
Hankel functions   100
  asymptotic formula   73
Harmonics   166
  cyclotron   (*see* cyclotron harmonics)
Harp, R. S.   174, 214, 215, 216
Hasegawa, A.   155, 161, 163, 213, 214
Heald, M. A.   210
Hellberg, M. A.   179, 215
Helmholtz equation   9
Herlofson, N.   10, 130, 132, 212
Herschberger, J.   213
Hexapolar effects   93
High-frequency conductivity   6 (*see also* equivalent conductivity tensor)
Hoh, F. C.   132, 213
Hollow cylindrical plasma   70 ff.
  theory   70 ff.
  comparison between theory and experiments   73 ff.
  asymmetry effects   *80* ff.
Horton, C. W.   165, 214

Index

Hot plasma slab-condenser system 47 ff.
Hsuan, H. C. 216
Huggins, R. W. 214
Hydrodynamic equations 4, 5
  cylindrical case 26
  tensorial generalization 28
  and damping of temperature resonances 53 ff, *150* ff.

Ichikawa, Y. 174, 215
Ichimaru, S. 214
Ichimiya, T. 174, 215
Ikegami, H. 173, 174, 215
Inductive behaviour (of a plasma) 47, *192*, 196
Inhomogeneous permittivity (*see* non-uniform permittivity)
Inhomogeneous plasma (*see* non-uniform plasma)
Ionic sheath (*see* sheath)
Ionosphere 184
Ivanov, A. A. 214

Jackson, E. A. 214
Johnston, T. W. 189, 215
Jump discontinuities *13*
  between two plasmas *154, 209*

Kaiser, T. R. 72, 130, 211
Keitel, G. 130, 213
Kerzar, B. 152, 213
Killian, T. G. 213
Kino, G. S. 16, 174, 211, 212, 213, 214, 215
Kononov, B. P. 214
Kroll, N. M. 214
Kubo, H. 214
Kuehl, H. H. 165, 214

Landau, L. D. 211
Landau damping 69, *150*
Landau solution 19
Landauer, G. 161, 214
Langmuir, I. 213
Laplace equation 9, 10, 71, 81, 176, 183, *201*
Laplace forces 108, *109*
Leavens, W. M. 211
Lepechinski, D. 174, 215
Leprince, P. 42, 216

Levitskii, S. M. 174, 215
Linear detection *31*
Linearization *17*, 19
Linearized theory *4*
Llewellyn-Jones, F. 210
Longitudinal plasma waves (or oscillations) 9, *17, 45, 134, 142,* 162
Lonngren, K. E. 216
Lorentz conductivity 6
Lorentz force 109
Lorentzian resonance profile *150*
Low, F. E. 14, 212
Lustig, C. D. 152, 213

Mackinson, R. E. B. 212
Macroscopic quantities *3, 4*
McCutchen, C. W. 215
Magistrelli, F. 177, 215
Malmberg, J. H. 149, 213
Mantei, T. D. 216
Maxwell's equations 4, *8, 95,* 118
Maxwell's surface equations *119*
Mayer, H. M. 174, 215
Measurement of electron plasma densities 41 ff., 79 ff., *179, 189*
Meltz, G. 165, 214
Messiaen, A. M. 31, 57, 70, 92, 94, 132, 134, 168, 174, 182, 191, 211, 212, 213, 214, 215, 216
Metallic resonance probe 173 ff.
  and static magnetic field *179*
Metallic surface 30, *160*
Miller, M. A. 216
Mitani, K. 214
Miyazaki, S. 173, 215
Miniaturization (of dielectric resonance probe) 185
Moments method *17* ff.
Momentum transfer equation 4, 5, 18, *21*
Monfort, J. L. 216
Montgomery, D. C. 210
Morse, P. 210
Mourier, G. 215
Mower, L. 212
Mueller, G. 174, 215
Multipole electrode configurations *40* ff.

Natural oscillation frequency (*see* oscillation frequency)

Neumann functions, asymptotic formula 73
Nickel, J. C.   132, 142, 212, 213
Nihoul, J. C.   154, 212, 213
Noise spectrum   *152* ff.
Nonlinear effects   *166* ff., *200*
  cold plasma at low power   *168* ff.
  plasma coated antenna   *200*
  strong effects   *171* ff.
  temperature resonances at low power   *166* ff.
Non-uniform cold plasma   8, 11, *13*, 16
  asymmetry   93
  cylindrical case   26
Non-uniform hot plasma   25 ff., *130* ff.
  scalar perturbed pressure   *132* ff.
  static axial magnetic field   *160* ff.
  tensorial perturbed pressure   25 ff.
  versus uniform cold plasma   *160*
Non-uniform permittivity   11
Nyquist's theorem   153

Obermann, C. R.   17, 212
O'Brien, B.   158, 214
Oscillation frequency   11
Oster, L.   211
Ozaki, H. T.   97, 211

Papas, C. H.   210
Parabolic density profile   *134* ff.
Parker, J. V.   132, 142, 158, 212, 213, 214
Pavkovitch, J.   174, 215
Pearson, G. A.   165, 214
Permittivity (equivalent permittivity of a plasma; *see* equivalent permittivity)
Perturbed quantities   4
Perturbed velocity   18
Perturbed pressure (*see* scalar electron pressure)
Perturbed pressure, tensorial   25 ff., 131, *146*
Peter, G.   174, 215
Petroff, I.   213
Pines, D.   214
Plasma coated antenna   *191* ff.
  and steady magnetic field   200
  non-linear behaviour   200
Plasma cylinder   10 ff., *72, 79, 130* ff.
Plasma frequency   6
Plasma slab-condenser system   43 ff., 174, 177
  cold plasma approximation   45 ff.
  collision effects   53 ff.
  equations   43 ff.
  equivalent electrical circuit   47
  experiments   57 ff.
  Vlasov's equation   63 ff.
  warm plasma   47 ff.
Plasmoids (high frequency)   *60* ff., 172
Platzman, P. M.   97, 211
Poincelot, P.   210
Pointu, A.   216
Polarization (high frequency)   29, 94
Pomeau, Y.   216

Q-factor   *150* ff.
Quadrupolar mode, cold plasma cylinder   93
  hot plasma cylinder   146
Quasi-static approximation   8 ff., 10, *134, 142*, 175, 182
  criterion   9
Quasi-static oscillations   13 (*see also* longitudinal plasma waves)

Rabben, H. H.   174, 215
Radiation conductance   195
Radiation damping   11, *149* ff.
Raether, M.   214
Reflector factor   34
Reflexion in free space   *31* ff.
Rem, J.   165, 214
Resonances and anti-resonances (general considerations)   *201* ff.
Resonance discharges (high frequency)   *60* ff., *171* ff., *200*
Resonance doublet   *81, 87*
Resonance probe, dielectric (*see* dielectric resonance probe)
  metallic (*see* metallic resonance probe)
Resonance spectrum (detailed) of a plasma column   *37* ff., *147* ff. (*see also* temperature resonances)
Resonant scattering   10 (*see also* cold plasma and temperature resonances)
Resonantly sustained plasmas   63, *171, 200*

Index

Roentgen, W. C. 128, 211
Roentgen–Eichenwald current 94, *117*, *128*, 129
Rolland, P. 174, 215
Romell, D. 130, 132, 212
Ron, A. 214
Rose, D. J. 212
Rostoker, N. 214
Ryutov, D. D. 214

Salpeter, E. E. 214
Satellite 185, 200
Scalar electron pressure 5, *132* ff.
Scattering from density fluctuations *169* ff.
Scattering of waves by plasma columns 30 ff., 70 ff., 92 ff., *130* ff.
Schmitt, H. J. 165, 213, 214
Second series (of temperature resonances) *148*, *157*, 159
Secondary resonances (*see* temperature resonances)
Self, S. A. 211, 216
Self consistent field 8
Sellberg, F. 152, 213
Series "limit" *141*, *148*
Shapiro, H. 211
Shashurin, I. P. 174, 215
Sheath 62, *174*, 175, 179
Sherby Harvie, R. S. R. 216
Shure, F. C. 63, 210
Simonen, T. C. 149, 216
Singular modes 13
Singular point 12
Sitenko, A. G. 210
Slade, D. M. 212
Slater, J. C. 210
Smith, D. F. 175, 215
Smith, F. H. C. 189, 215
Sommerfeld, A. 210
Spalter, J. 211
Spitzer, L. Jr. 210
Static magnetic induction 7
Stepanov, K. N. 210
Stern, R. A. 166, 169, 213, 214
Stix, T. H. 89, 165, 210, 214
Sturrock, P. A. 14, 212
Subsidiary resonances (*see* temperature resonances)

Taillet, J. 60, 174, 211, 215

Takayama, K. 173, 215
Tanaka, S. 214
TE mode 95, 97, *99* ff., *105*, *110* ff., *112* ff., *121*
Temperature resonances 37, *130*, *132* ff.
  asymmetry effects *147*, *154* ff.
  axial magnetic field *160* ff.
  axial propagation *158* ff.
  damping of *149* ff.
  detailed structure of spectrum *147* ff.
  historical introduction *130* ff.
  in a plasma slab-condenser *51* ff.
  phenomenological picture *140* ff.
  second series of resonances *148*, *157*
  with gases other than Hg *131*, *151*
Tensorial generalization of hydrodynamic equations 28
Tensorial permittivity (*see* equivalent permittivity)
Tensorial perturbed pressure *25* ff.
Thomassen, K. I. 42, 212
Thompson, W. B. 26, 29, 210
Tidman, D. A. 210
Time-dependence 4
TM mode 95, 97, *105*, *114* ff., 121
Tonks, L. 10, 130, 212
Tonks–Dattner resonances *132* (*see* temperature resonances)
Tonks–Langmuir density profile 132, 140, *142* ff., 143
Transverse electromagnetic waves 134
Trivelpiece, A. W. 214
Tzoar, N. 169, 214

Uniform cold plasma column 10, 72, 78, 79

Vandenplas, P. E. 17, 23, 31, 43, 57, 70, 92, 94, 131, 132, 134, 154, 168, 182, 191, 210, 211, 212, 213, 214, 215, 216
Van Hoven, G. 149, 216
Van Kampen, N. G. 20, 212
Variational procedures *13* ff.
Velocity (perturbed) 18
Vlasov equation (*see* Boltzmann–Vlasov equation)
von Engel, A. 211
von Gierke, G. 215

Waletzko, J. A. 215
Wasserab, T. 211
W B K approximation 141
Weenink, M. P. H. 165, 214
Weibel, E. S. 216
Weiss, H. H. 214

Weissglas, P. 132, 142, 152, 210, 213
Wharton, C. B. 149, 210, 213
Whittaker 104
Willis, J. 213
Wimmel, H. 174, 215